Airplane Design

Part V: Component Weight Estimation

Dr. Jan Roskam
Ackers Distinguished Professor of Aerospace Engineering
The University of Kansas, Lawrence

2003

1440 Wakarusa Drive, Suite 500 • Lawrence, Kansas 66049, U.S.A.

PUBLISHED BY
Design, Analysis and Research Corporation (*DARcorporation*)
1440 Wakarusa Drive, Suite 500
Lawrence, Kansas 66049
U.S.A.
Phone: (785) 832-0434
Fax: (785) 832-0524
e-mail: info@darcorp.com
http://www.darcorp.com

Library of Congress Catalog Card Number: 97-68580

ISBN 1-884885-50-0

In all countries, sold and distributed by
Design, Analysis and Research Corporation
1440 Wakarusa Drive, Suite 500
Lawrence, Kansas 66049
U.S.A.

Printed in the United States of America
First Printing, 1985
Second Printing, 1999
Third Printing, 2003

TABLE OF CONTENTS

TABLE OF SYMBOLS

Symbol	Definition	Dimension
A	wing aspect ratio	-----
$A_{h,v,c}$	Hor. tail, Vert. tail or Canard aspect ratio	-----
A_{inl}	Inlet capture area per inlet	ft^2
A_g	constant in Eqn.(5.42) and Table 5.1	
b	wing span	ft
$b_{h,v,c}$	Hor. tail, Vert. tail or Canard span	ft
B_g	constant in Eqn.(5.42) and Table 5.1	
$\bar{c}$	wing mean geometric chord	ft
$\bar{c}_{h,v,c}$	mean geometric chord of hor. tail, vert. tail or canard	ft
C_g	constant in Eqn.(5.42) and Table 5.1	
C_D	Drag coefficient	-----
C_L	Lift coefficient	-----
C_{L_α}	Airplane lift-curve slope	rad^{-1}
C_N	Normal force coefficient	-----
D_g	constant in Eqn.(5.42) and Table 5.1	
D_p	Propeller diameter	ft
$e = (b + L)/2$	Used in inertia calcs.	ft
FAR	Federal Air Regulation	-----
g	acceleration of gravity	ft/sec^2
GW	Flight design gross wht	lbs
h	altitude	ft
h_f	maximum fuselage height	ft
int	fraction of fuel tanks which are integral	-----
I	Moment of inertia	$slugs/ft^3$

K = constant as defined in equations below:

K_{api} (7.32)	K_b (6.34)	K_{bc} (7.48)	K_{buf} (7.44)
K_c (4.6)	K_d (6.9) and (6.10) but note that values differ		K_{ec} (6.23)
K_f (5.27)	K_{fcf} (7.9)	K_{gr} (5.42)	K_h (5.19)
K_{inl} (5.26) and (5.28)		K_{lav} (7.44)	K_m (6.9)
K_n (5.29)	K_{osc} (6.38)	K_p (6.2)	K_{pg} (6.4)
$K_{prop1\ or\ 2}$ (6.13) or (6.14)			K_r (6.11)
K_s (6.12)	K_{thr} (6.6)	K_v (5.20)	K_w (5.9), (5.10)
K_{st} (7.46)			

K_{fsp}	specific weight of fuel	lbs/gal
K_g	Gust alleviation factor, see Eqn.(4.16)	
l_f	length of fuselage	ft
l_{f-n}	length of fuselage minus nacelle	ft
$l_{h,v,c}$	Distance from wing 1/4c to $1/4c_{h,v,c}$	ft
l_{inl}	inlet length from lip to compressor face	ft
l_{pax}	length of passenger cabin	ft
$l_{s_{m\ or\ n}}$	shock strut length for main gear or for nose gear	ft
L	Overall airplane length	ft
L_d	inlet duct length	ft
L_r	ramp length	ft
M	Mach number	
M_{ff}	Mission fuel fraction (M_{ff}= End weight/Begin weight)	none
n	Load factor	-----
N	Number of (see subscript)	-----

p_{max}	maximum fusel. perimeter	ft
P_c	design ult. cabin press.	psi
P_{TO}	required take-off power	hp
P_2	maximum static pressure at engine compressor face	psi
$\bar{q}$	dynamic pressure	psf
R	Range	nm or m
$R_{x,y,z}$	Radius of inertia about x,y,z axis respectively	ft
$\bar{R}_{x,y,z}$	Non-dimensional radius of inertia about x,y,z axiz resp.	-----
S	Wing area	ft^2
S_{cs}	Total control surface area	ft^2
S_{ff}	freight floor area	ft^2
S_{fgs}	Fuselage gross shell area	ft^2
$S_{h,v,c}$	Hor., Vert. or Can. area	ft^2
S_r	Rudder area	ft^2
SHP	Shaft horsepower	hp
t/c	thickness ratio	-----
t_r	maximum root thickness	ft
U_{de}	Derived gust velocity	fps
V	True airspeed	mph, fps, kts
V_A	Design maneuvering speed	KEAS
V_B	Design speed for maximum gust intensity	KEAS
V_C	Design cruise speed	KEAS
V_D	Design dive speed	KEAS
V_H	Maximum level speed at at sealevel	KEAS

V_{pax}	Volume of passenger cabin	ft^3
$V_{pax+cargo}$	Vol. of pass. and cargo	ft^3
V_{pr}	Pressurized volume	ft^3
V_S, V_{S_1}	+1g stall speed	KEAS
w_f	maximum fuselage width	ft
W	Weight	lbs
W_i	Weight of component i	lbs
x	distance from some ref.	ft
x_i	distance from some ref. of component i	ft
z_h	Distance from vert.tail root to where h.t. is mounted on the v.t.	ft

Greek Symbols
=============

α	angle of attack of airplane	rad.
ε	downwash angle at h.t.	rad.
ρ	air density	$slugs/ft^3$
λ	wing taper ratio	-----
$\lambda_{h,v,c}$	taper ratio for hor. tail, vert. tail or canard	-----
μ_g	airplane mass ratio, see Eqn.(4.17)	
Λ_n	sweep angle at n^{th} chord station	

Subscripts
==========

ai	air induction
api	airconditioning, pressurization, de-icing and anti-icing system
apsi	accessory drives, powerplant controls, starting and ignition system
apu	auxiliary power unit
arm	armament
aux	auxiliary
bal	ballast
bc	baggage and cargo handling equipment
bl	blades
c	canard
cc	cabin crew
cg	center of gravity
cr	crew
crew	crew
C	Cruise

D	Dive
e	engines (all!)
ec	engine controls
els	electrical system
emp	empennage
eng	engine (one only)
ess	engine starting system
etc	etcetera (please pronounce as eTcetera and *not* as eKcetera)
E	Empty
f	fuselage
fc	flight control system
fd	fuel dumping system
fdc	flight deck crew
feq	fixed equipment
fl.boat	flying boat
fs	fuel system
fti	flight test instrumentation
fur	furnishings
F	Mission fuel
g	landing gear
glw	guns, launchers and weapons provisions
h	horizontal tail
hps	hydraulic and pneumatic system
H	maximum level flight at sealevel
i	instrumentation
iae	instrumentation, avionics and electronics
inflref	in-flight refuelling system
inl	inlet(s)
lim	limit
L	Landing (subscript to W)
L	maximum dive (subscript to V)
LE	Leading Edge
m	maximum
max	maximum
MZF	Maximum zero fuel
n	nacelle
neg	negative
ops	operational items
osc	oil system and oil cooler
ox	oxygen system
pax	passengers
p	propellers (subscript to N)
p	propulsion system (subscript to W)
pc	propeller controls
pos	positive
prop	propeller
pt	paint
pwr	powerplant
PL	Payload

ramp	ramp
sprchr	supercharger
struct	structure
supp	bladder support structure
t	fuel tanks
tfo	trapped fuel and oil
tr	thrust reverser system
troop	troop(s)
TO	Take-off
ult	ultimate
ult.l.	ultimate landing
v	vertical tail
w	wing
wb	wing + body
wi	water injection system
xx, yy, zz	about x-, y-, z-axis respectively

Acronyms
========

APU	Auxiliary power unit
C.G., c.g.	Center of gravity
OWE	Operating weight empty
shp	shaft horse power
TBP	Turboprop

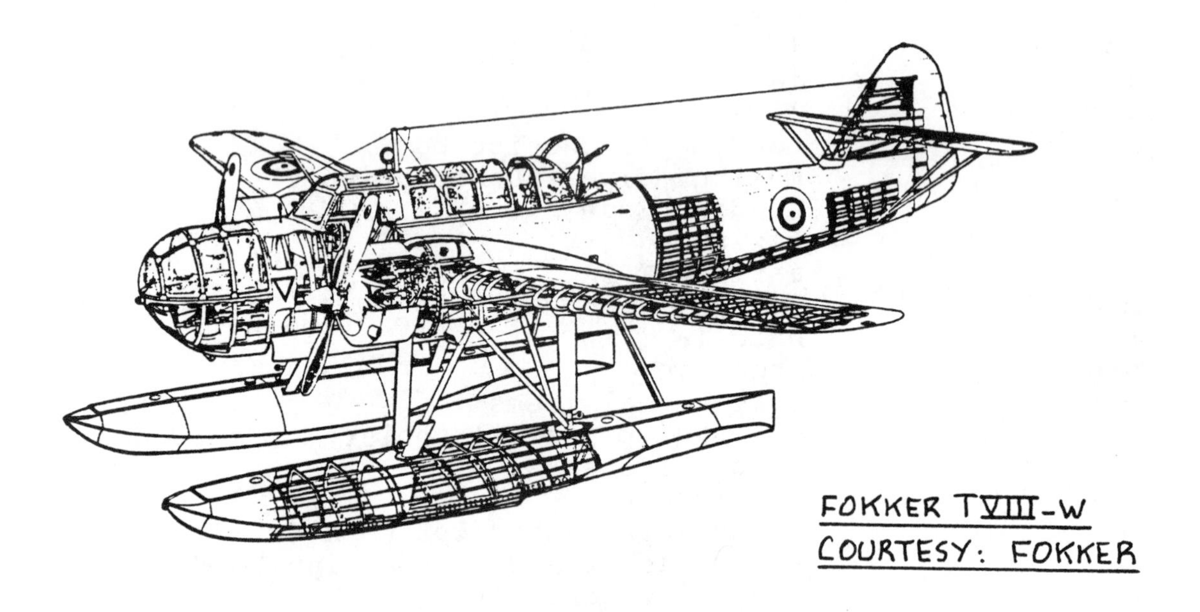

FOKKER TVIII-W
COURTESY: FOKKER

ACKNOWLEDGEMENT

Writing a book on airplane weight estimation is impossible without the supply of a large amount of data. The author is grateful to the following companies for supplying the weight data, the weight manuals and the weight estimating procedures which made the book what it is:

Beech Aircraft Corporation
Boeing Commercial Airplane Company
Canadair
Cessna Aircraft Company
DeHavilland Aircraft Company of Canada
Gates Learjet Corporation
Gulfstream Aerospace Corporation
Lockheed Aircraft Corporation
McDonnell Douglas Corporation
NASA, Ames Research Center
Rinaldo Piaggio S.p.A.
Rockwell International
Royal Netherlands Aircraft Factory, Fokker
SIAI Marchetti S.p.A.

A significant amount of airplane design information has been accumulated by the author over many years from the following magazines:

Interavia (Swiss, monthly)
Flight International (British, weekly)
Business and Commercial Aviation (USA, monthly)
Aviation Week and Space Technology (USA, weekly)
Journal of Aircraft (USA, AIAA, monthly)

The author wishes to acknowledge the important role played by these magazines in his own development as an aeronautical engineer. Aeronautical engineering students and graduates should read these magazines regularly.

FOKKER F.27 FRIENDSHIP

COURTESY: FOKKER

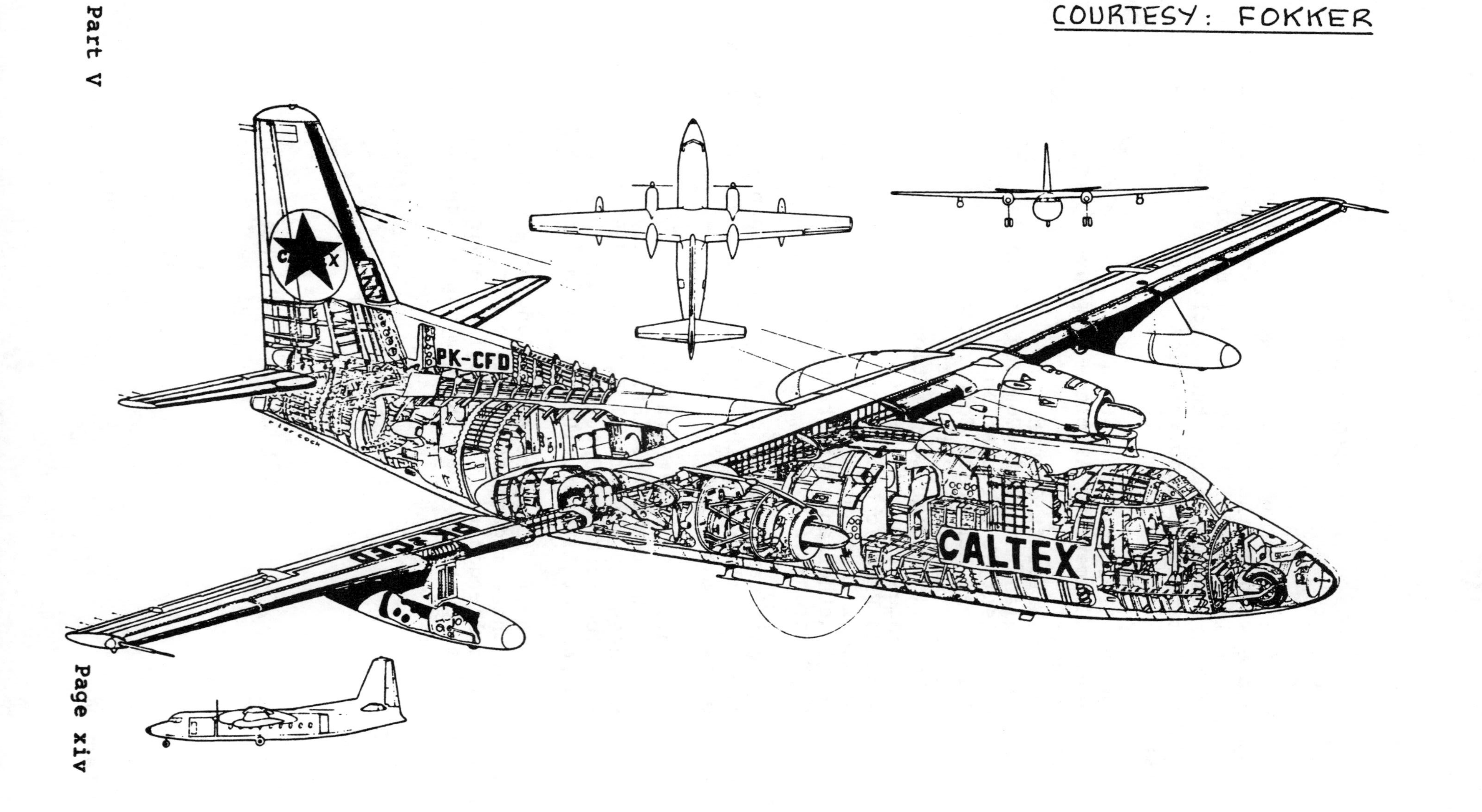

1. INTRODUCTION

The purpose of this series of books on Airplane Design is to familiarize aerospace engineering students with the design methodology and design decision making involved in the process of designing airplanes.

The series of books is organized as follows:

PART I:	PRELIMINARY SIZING OF AIRPLANES
PART II:	PRELIMINARY CONFIGURATION DESIGN AND INTEGRATION OF THE PROPULSION SYSTEM
PART III:	LAYOUT DESIGN OF COCKPIT, FUSELAGE, WING AND EMPENNAGE: CUTAWAYS AND INBOARD PROFILES
PART IV:	LAYOUT DESIGN OF LANDING GEAR AND SYSTEMS
PART V:	COMPONENT WEIGHT ESTIMATION
PART VI:	PRELIMINARY CALCULATION OF AERODYNAMIC, THRUST AND POWER CHARACTERISTICS
PART VII:	DETERMINATION OF STABILITY, CONTROL AND PERFORMANCE CHARACTERISTICS: FAR AND MILITARY REQUIREMENTS
PART VIII:	AIRPLANE COST ESTIMATION: DESIGN, DEVELOPMENT, MANUFACTURING AND OPERATING

The purpose of PART V is to present methods for estimating airplane component weights and airplane inertias during airplane preliminary design.

Two methods are presented: they are called the Class I and the Class II method respectively.

The Class I method relies on the estimation of a percentage of the flight design gross weight (= take-off weight for most airplanes) of major airplane components. These percentages are obtained from actual weight data for existing airplanes. The usual procedure is to average these percentages for a number of airplanes similar to the one being designed. These averaged percentages are multiplied by the take-off weight to obtain a first estimate of the weight of each major component.

The method can be used with minimal knowledge about the airplane being designed and requires very little engineering work. However, the accuracy of this method is limited. It should be used only in association with preliminary design sequence I as outlined in Part II (See Step 10, p.15).

Chapter 2 presents the Class I method for estimating

airplane component weights in the form of a step-by-step precedure. Three example applications are also given.

Chapter 3 presents a Class I method for estimating airplane moments of inertia. Example applications are also given.

Class II methods are based on weight equations for more detailed airplane components and groupings. These equations have a statistical basis. They do allow the designer to account for fairly detailed configuration design parameters. To use this method it is necessary to have a V-n diagram, a preliminary structural arrangement and to have decided on all systems which are needed for the operation of the airplane under study.

The Class II method should be used in conjunction with preliminary design sequence II as outlined in Part II (See Step 21, p.19).

Chapter 4 presents the Class II method for estimating airplane component weights in the form of a step-by-step procedure. A method for construction of a V-n diagram is included. Example applications are given.

As part of the Class II weight estimation procedure the airplane empty weight is split into three major groupings:

1. Structure weight
2. Powerplant weight
3. Fixed equipment weight

Chapters 5, 6 and 7 present the detailed methodologies used in determining the component weights within each of these three groupings.

Chapter 8 contains data and methods for rapidly determining the c.g. location of individual components.

A Class II method for performing a weight and balance analysis is discussed in Chapter 9.

Chapter 10 presents a Class II method for computing airplane moments and products of inertia.

Appendix A contains a data base for airplane component weights and weight fractions.

Appendix B contains a data base for non-dimensional radii of gyration for airplanes.

2. CLASS I METHOD FOR ESTIMATING AIRPLANE COMPONENT WEIGHTS

The purpose of this chapter is to provide a methodology for rapidly estimating airplane component weights. The emphasis is on rapid and on spending as few engineering manhours as possible. Methods which fit meet these objectives are referred to as Class I methods. They are used in conjunction with the first stage in the preliminary design process, the one referred to as 'p.d. sequence I' in Part II (See Step 10, p.15).

The Class I weight estimating method relies on the assumption, that within each airplane category it is possible to express the weight of major airplane components (or groups) as a simple fraction of one of the following weights:

1. Gross take-off weight, W_{TO}

2. Flight design gross weight, GW

3. Empty weight, W_E

The reader is already familiar with the definition of W_{TO} and W_E. The flight design gross weight, GW is that weight at which the airplane can sustain its design ultimate load factor, n_{ult}. For civil airplanes GW and W_{TO} are often the same, although there are exceptions. For military airplanes GW and W_{TO} are frequently quite different.

In this book, all component weight fractions are given relative to the flight design gross weight, GW. In the component weight and weight fraction data presented in Appendix A, both GW and W_{TO} are listed for all airplanes for which data are presented.

Since W_{TO} is known from the preliminary sizing work described in Part I, the value of GW can be established.

The weight of any major airplane component or group can now be found rapidly through multiplication of GW by an appropriate weight fraction. For this reason, the Class I weight method is also referred to as the 'weight fraction' method.

Section 2.1 presents a step-by-step procedure for using weight fractions to estimate the component weight breakdown of airplanes.

Section 2.2 presents example applications to three airplanes.

2.1 A METHOD FOR ESTIMATING AIRPLANE COMPONENT WEIGHTS WITH WEIGHT FRACTIONS

In this section the Class I method for estimating airplane component weights is presented in the form of a step-by-step procedure.

Step 1: List the following overall weight values for the airplane:

1. Gross take-off weight, W_{TO}
2. Empty weight, W_E
3. Mission Fuel Weight, W_F
4. Payload weight, W_{PL}
5. Crew weight, W_{crew}
6. Trapped fuel and oil weight, W_{tfo}
7. Flight design gross weight, GW

Weight items 1-6 are already known from the preliminary sizing process described in Part I (See Chapter 2).

For most airplanes, W_{TO} and GW are the same. In the case of many military airplanes there is a difference. Appendix A contains tables with airplane weight data on basis of which a decision can be made about the ratio between W_{TO} and GW. Sometimes the mission specification will include this information.

Step 2: Proceed to Appendix A and determine which airplane category best fits the airplane which is being designed. Identify those airplanes which will be used in estimating the weight fractions for the airplane which is being designed.

Step 3: Make a list of the significant airplane components for which weights need to be estimated. This list will vary some from one airplane type to the other. In many cases certain weight items are already specified in the mission specification.

A typical Class I component weight list contains the following items:

I. Structure Weight, W_{struct}

1. Wing
2. Empennage
   2.1 Horizontal tail and/or canard
   2.2 Vertical tail and/or canard
3. Fuselage (and/or tailbooms)
4. Nacelles
5. Landing gear
   5.1 Nose gear
   5.2 Main gear
   5.3 Tail gear
   5.4 Outrigger gear
   5.5 Floats

II. Powerplant Weight, W_{pwr}

1. Engine(s), this may include afterburners or thrust reversers
2. Air induction system
3. Propeller(s)
4. Fuel system
5. Propulsion system

III. Fixed Equipment Weight, W_{feq}

1. Flight control system
2. Hydraulic and pneumatic system
3. Electrical system
4. Instrumentation, avionics and electronics
5. Air conditioning, pressurization, anti-icing and de-icing system
6. Oxygen system
7. Auxiliary power unit
8. Furnishings
9. Baggage and cargo handling equipment
10. Operational items

11. Armament
12. Guns, launchers and weapons provisions
13. Flight test instrumentation
14. Auxiliary gear
15. Ballast
16. Paint
17. Other weight items not listed above

Consult the mission specification as well as the appropriate tables in Appendix A for any weight items not listed above.

The airplane empty weight, W_E is expressed as:

$$W_E = W_{struct} + W_{pwr} + W_{feq} \quad (2.1)$$

Whether or not it is necessary to split weight groupings II and III in as many components as listed above depends on the expected effect of these components on the accuracy of the airplane c.g. location.

Use as much detail as necessary for realism in the Class I weight and balance analysis of Chapter 10, Part II.

Step 4: From the appropriate Table(s) in Appendix A decide on the weight fractions to be used.

Frequently it will be sufficient to use average fraction values obtained from a number of airplanes with missions not too much different from the mission of the airplane being designed. The reader should familiarize himself with what the airplanes for which weight fraction data are available, look like and what their missions were. This can be done by referring to Jane's All the World Aircraft (Ref. 8). Jane's contains an index identifying which issue of Jane's contains descriptions of certain types of airplanes.

It is of great importance to observe whether or not:

1. an airplane has a strutted (braced) wing
2. an airplane is pressurized
3. the landing gear is mounted on the fuselage or on the wing
4. the engines are mounted on the wing or fuselage

The reader should note, that most weight and weight fraction data in Appendix A are for airplanes with largely aluminum primary structures. If the airplane being designed will have to contain a significant amount of primary structure made from composites, from lithium-aluminum or from other materials, it will be necessary to modify the weight fractions. Table 2.16, p.48, Part I may be useful in this regard.

After thus 'massaging' the weight fraction data, list the weight fractions to be used. Make careful notes of reasons why specific fractions were selected.

Step 5: Multiply the selected weight fractions by the GW value of Step 1 and list all significant airplane component weights.

The Class I component weight data thus obtained are used in the Class I weight and balance analysis described in Chapter 10 of Part II.

To illustrate the use of this procedure, three examples are presented in Section 2.2.

Step 6: Document the decisions made under Steps 1 through 5 in a brief, descriptive report.

2.2 EXAMPLE APPLICATIONS

In this section, three example applications of the Class I component weight estimating method will be discussed:

2.2.1 Twin Engine Propeller Driven Airplane: Selene
2.2.2 Jet Transport: Ourania
2.2.3 Fighter: Eris

2.2.1 Twin Engine Propeller Driven Airplane

Step 1: Overall weight values for this airplane were determined as a result of the preliminary sizing performed in Part I. These weight values are summarized in sub-sub-section 3.7.2.6, Part I, p.178:

W_{TO} = 7,900 lbs $\quad$ W_E = 4,900 lbs

W_F = 1,706 lbs $\quad$ W_{PL} = 1,250 lbs (Part I, p.49)

W_{tfo} = 44 lbs makes up the balance.

The crew weight is included in the payload of this airplane. It will be assumed that GW = W_{TO}. This is consistent with the data in Tables A3.1 and A3.2.

For easy reference the airplane will be referred to as the Selene, the name of the Greek Moon Goddess.

Step 2: Tables A3.1 and A3.2 contain component weight data for airplanes in the same category as the Selene. Specifically, the following airplanes have comparable sizes and missions: Cessna 310C, Beech 65 Queen Air, Cessna 404-3 and Cessna 414A.

Step 3: For reasons of brevity, only the following component weights are considered:

Wing	Empennage	Fuselage	Nacelles
Landing Gear	Power Plant	Fixed Eqpmt	

Step 4: The following table lists the pertinent weight fractions and their averaged values. Because the intent is to apply conventional metal construction methods to the Selene there is no reason to alter the averaged weight fractions.

	Beech 65 QA	Cessna 310C	Cessna 404-3	Cessna 414A	Selene Average
Pwr Plt/GW	0.219	0.259	0.194	0.206	0.220
Fix Eqp/GW	0.123	0.103	0.134	0.167	0.132
Empty Wht/GW	0.638	0.628	0.596	0.665	0.631
Wing Grp/GW	0.091	0.094	0.102	0.094	0.095
Emp. Grp/GW	0.021	0.024	0.022	0.024	0.023
Fus. Grp/GW	0.082	0.066	0.073	0.100	0.080
Nac. Grp/GW	0.039	0.027	0.034	0.029	0.032
Gear Grp/GW	0.060	0.054	0.038	0.045	0.049

Note that the ratio of W_E/GW which follows from the preliminary sizing, is 4,900/7,900 = 0.62. This is close to the average value of 0.631 in the above tabulation.

Step 5: Using the averaged weight fractions from Step 4, the following preliminary component weight summary can be determined:

Component	First weight estimate lbs	Adjustment lbs	Selene Class I weight (alum.) lbs	Selene Class I weight (compos.) lbs
Wing	751	-13	738	627
Empennage	182	- 3	179	152
Fuselage	632	-11	621	528
Nacelles	253	- 4	249	212
Landing Gear	387	- 7	380	380
Power Plant	1,738	-30	1,708	1,708
Fixed Eqp.	1,043	-18	1,025	1,025
Empty Wht	4,986	-86	4,900	4,632
Payload			1,250	1,250
Fuel			1,706	1,706
Trapped fuel and oil			44	44
Take-off Gross Weight			7,900	7,632

When the numbers in the first column are added, they yield an empty weight of 4,986 lbs instead of the desired 4,900 lbs. The difference is due to round-off errors in the weight fractions used. It is best to 'distribute' this difference over all items in proportion to their component weight value listed in the first column.

For example, the wing adjustment number is arrived at by multiplying 86 lbs by 751/4986.

It is quite possible that in other airplanes the adjustment will turn out to be positive instead of negative.

If the judgement is made to manufacture the Selene with composites as the primary structural materials significant weight savings can be obtained. A conservative assumption is to apply a 15 percent weight reduction to wing, empennage, fuselage and nacelles. The resulting weights are also shown in the Class I weight tabulation. Note the reduction in empty weight of 268 lbs. Using the weight sensitivity $\partial W_{TO}/\partial W_E = 1.66$ as computed in sub-sub-section 2.7.3.1 in Part I, an overall reduction in W_{TO} of 1.66x268 = 545 lbs can be achieved.

The designer has the obvious choice to fly the same mission with (545 - 268) = 277 lbs less fuel or to simply add the 545 lbs to the useful load of the Selene.

The component weight values in the column labelled: 'Class I weight (alum.)' are those to be used in the Class I weight and balance analysis of the Selene. This corresponds to Step 10 as outlined in Chapter 2, Part II. The Class I weight and balance analysis for the Selene is carried out in Chapter 10 of Part II (See pp. 246-250).

Step 6: To save space, this step has been omitted.

2.2.2 Jet Transport

Step 1: Overall weight values for this airplane were determined as a result of the preliminary sizing performed in Part I. These weight values are summarized in sub-sub-section 3.7.3.6, Part I, p.183:

W_{TO} = 127,000 lbs W_E = 68,450 lbs

W_F = 25,850 lbs W_{PL} = 30,750 lbs (Part I, p.54)

W_{tfo} = 925 lbs W_{crew} = 1,025 lbs (Part I, p.58)

It will be assumed that $GW = W_{TO}$ for this airplane.

This is consistent with the data in Tables A7.1 through A7.5.

For easy reference the airplane will be referred to as the Ourania, the name of the Greek Muse of Astronomy.

Step 2: Tables A7.1 through A7.5 contain component weight data for airplanes in the same category as the Ourania. Specifically the following airplanes have comparable sizes and missions: McDonnell-Douglas DC-9-30 and MD-80, Boeing 737-200 and 727-100.

Step 3: For reasons of brevity, only the following component weights are considered:

Wing	Empennage	Fuselage	Nacelles
Landing Gear	Power Plant	Fixed Eqpmt	

Step 4: The following table lists the pertinent weight fractions and their averaged values. Because the intent is to apply conventional metal construction methods to the Ourania, there is no reason to alter the averaged weight fractions.

	McDonnell-Douglas DC-9-30	MD-80	Boeing 737-200	727-100	Ourania Average
Pwr Plt/GW	0.076	0.079	0.071	0.078	0.076
Fix Eqp/GW	0.175	0.182	0.129	0.133	0.155
Empty Wht/GW	0.538	0.564	0.521	0.552	0.544
Wing Grp/GW	0.106	0.111	0.092	0.111	0.105
Emp. Grp/GW	0.026	0.024	0.024	0.026	0.025
Fus. Grp/GW	0.103	0.115	0.105	0.111	0.109
Nac. Grp/GW	0.013	0.015	0.012	0.024	0.016
Gear Grp/GW	0.039	0.038	0.038	0.045	0.040

Note that the ratio of W_E/GW which follows from the preliminary sizing, is 68,450/127,000 = 0.539. This is close to the average value of 0.544 in the above tabulation.

Step 5: Using the averaged weight fractions just determined, the following preliminary component weight summary can be determined:

Component	First weight estimate lbs	Adjustment lbs	Ourania Class I weight (alum.) lbs	Ourania Class I weight (li/alum.) lbs
Wing	13,335	+329	13,664	12,298
Empennage	3,175	+ 78	3,253	2,928
Fuselage	13,843	+341	14,184	12,766
Nacelles	2,032	+ 50	2,082	1,874
Landing Gear	5,080	+125	5,205	5,205
Power Plant	9,652	+239	9,891	9,891
Fixed Eqp.	19,685	+486	20,171	20,171
Empty Wht	66,802	+1,648	68,450	65,133
Payload			30,750	30,750
Crew			1,025	1,025
Fuel			25,850	25,850
Trapped fuel and oil			925	925
Take-off Gross Weight			127,000	123,683

When the numbers in the first column are added, they yield an empty weight of 66,802 lbs instead of the desired 68,450 lbs. The difference is due to round-off errors in the weight fractions used. It is best to 'distribute' this difference over all items in proportion to their component weight values listed in the first column.

For example, the wing adjustment number is arrived at by multiplying 1,648 lbs by 13,335/66,802. When so doing, the sum of the adjusted component weights is still 41 lbs shy of the desired goal. That new difference is then redistributed in the same manner.

It will be noted that the adjustments here are positive whereas for the light twin they were negative. It all depends on the weight fraction roundoffs, how this comes out.

If the judgement is made to manufacture the Ourania with lithium/aluminum as the primary structural material, sigificant weight savings can be obtained. A reasonable assumption is to apply a 10 percent weight reduction to wing, empennage, fuselage and nacelles. The resulting weights are also shown in the Class I weight tabulation. Note the reduction in empty weight of 3,317 lbs. Using the weight sensitivity $\partial W_{TO}/\partial W_E = 1.93$ as computed in

sub-sub-section 2.7.3.2 in Part I, an overall reduction in W_{TO} of 1.93x3,317 = 6,402 lbs can be achieved.

The designer has the obvious choice to fly the same mission with (6,402 - 3,317) = 3,085 lbs less fuel or to add the 6,402 lbs to the useful load of the Ourania.

The component weight values in the column labelled: 'Class I weight (alum.)' are those to be used in the Class I weight and balance analysis of the Ourania. This corresponds to Step 10 as outlined in Chapter 2, Part II. The Class I weight and balance analysis of the Ourania is carried out in Chapter 10 of Part II (See pp. 250-254.

Step 6: To save space, this step is omitted.

2.2.3 Fighter

Step 1: Overall weight values for this airplane were determined as a result of the preliminary sizing performed in Part I. These weight values are summarized in sub-sub-section 3.7.4.5, Part I, p.191:

W_{TO} = 64,500 lbs W_E = 33,500 lbs

W_F = 18,500 lbs W_{PL} = 12,000 lbs (Part I, p.60)

W_{tfo} = 300 lbs W_{crew} = 200 lbs (Part I, p.66)

It will be assumed that GW = $0.95W_{TO}$ for this airplane. This is consistent with the data in Tables A9.1 through A9.6.

For easy reference the airplane will be referred to as the Eris, the name of the Greek Goddess of War.

When looking up the actual bomb weight for a nominal 500 lbs bomb, it will be discovered that this weight is 531 lbs and not 500 lbs. That is a difference of 20x31 = 620lbs. On the other hand, the normal ammunition for the standard GAU-8A gun drum weighs 1,785 and not 2,000 lbs. The difference is -215 lbs. The actual payload is therefore 405 lbs more than originally planned.

<u>Step 2:</u> Tables A9.1 through A9.6 contain component weight data for airplanes in the same category as the Eris. Specifically the following airplanes have comparable sizes and missions: Republic F105B, Vought F8U, and Grumman A2F.

<u>Step 3:</u> For reasons of brevity only the following component weights are considered:

Wing	Empennage	Fuselage	Eng. Sect.
Landing Gear	Power Plant	Fixed Eqpmt	

<u>Step 4:</u> The following table lists the pertinent weight fractions and their averaged values. Since Eris will be made from conventional aluminum materials, there is no reason to alter the averaged weight fractions.

	Republic F105B	Vought F8U	Grumman A2F(A6)	Eris Average
Pwr Plt/GW	0.246	0.257	0.162	0.222
Fix Eqp/GW	0.155	0.135	0.159	0.150
Empty Wht/GW	0.797	0.722	0.651	0.723
Wing Grp/GW	0.109	0.135	0.136	0.127
Emp. Grp/GW	0.031	0.034	0.024	0.030
Fus. Grp/GW	0.187	0.126	0.102	0.138
Eng.Sect./GW	0.003	0.003	0.002	0.003
Gear Grp/GW	0.059	0.031	0.067	0.052
Engine(s)/GW	0.197	0.197	0.115	0.170
$n_{ult.}$	13	9.6	N.A	Use: 12
GW/W_{TO}	0.92	0.79	1.0	Use: 0.95

Note: all fraction data were based on GW without external stores!

Note that the ratio of W_E/GW which follows from the preliminary sizing, is 33,500/54,500 = 0.615. This is lower than the average value of 0.723 in the above tabulation. The reason is that the data base is for older fighters, two of which are USN fighters. Also note the large value $n_{ult.}$ for the F105B.

<u>Step 5:</u> Using the averaged weight fractions just determined, the following preliminary component weight summary can be determined:

Component	First weight estimate lbs	Adjustment lbs	Eris Class I weight (alum.) lbs
Wing	6,922	-160	6,762
Empennage	1,635	- 38	1,597
Fuselage	7,521	-174	7,347
Eng.Sect.	164	- 4	160
Landing Gear	2,834	- 66	2,768
Power plant	12,099 predicted from fraction data		
Engines	9,265 predicted from fraction data		
Engines	6,000 actual for F404's with A/B		
Engines			6,000
Eng.Sect.		12,099-9,265 =	2,834
Fix.Eqpmt	8,175 predicted from fraction data		
Ammo	2,000 (original estim.)		
Fix.Eqpmt-Ammo	6,175	-143 = 6,032	
GAU-8A Gun (Actual weight)			2,014
Fix.Eqpmt-Gun			4,018
Empty Wht	39,350	-585	33,500
Pilot			200
Payload: ammo			1,785
: bombs			10,620
Trapped fuel and oil			300
Fuel			18,500
Take-off Gross Weight			64,905

When the numbers in the first column are added, they yield an empty weight of 39,350 lbs instead of the desired 33,500 lbs., obtained from preliminary sizing. The difference is due to:

1. 2,000 lbs of ammo are included.
2. 3,265 lbs because of the much more favorable engine weight (9,265-6,000).
3. the remaining -585 lbs is due to round-off errors in the weight fractions.

The -585 lbs is distributed over all items which are computed with the weight fractions. This distribution is done in proportion to their component weight values in the first column.

For example, the wing adjustment number is arrived at by multiplying -585 lbs by 6,922/25,251*.

Note:

*25,251 = 6,922 + 1,635 + 7,521 + 164 + 2,834 + 6,175

The component weight values in the last column are those to be used in the Class I weight and balance analysis of the Eris. This corresponds to Step 10 as outlined in Chapter 2, Part II. The Class I weight and balance analysis of the Eris is carried out in Chapter 10 of Part II (See pp. 254-258).

Step 6: To save space, this step is omitted.

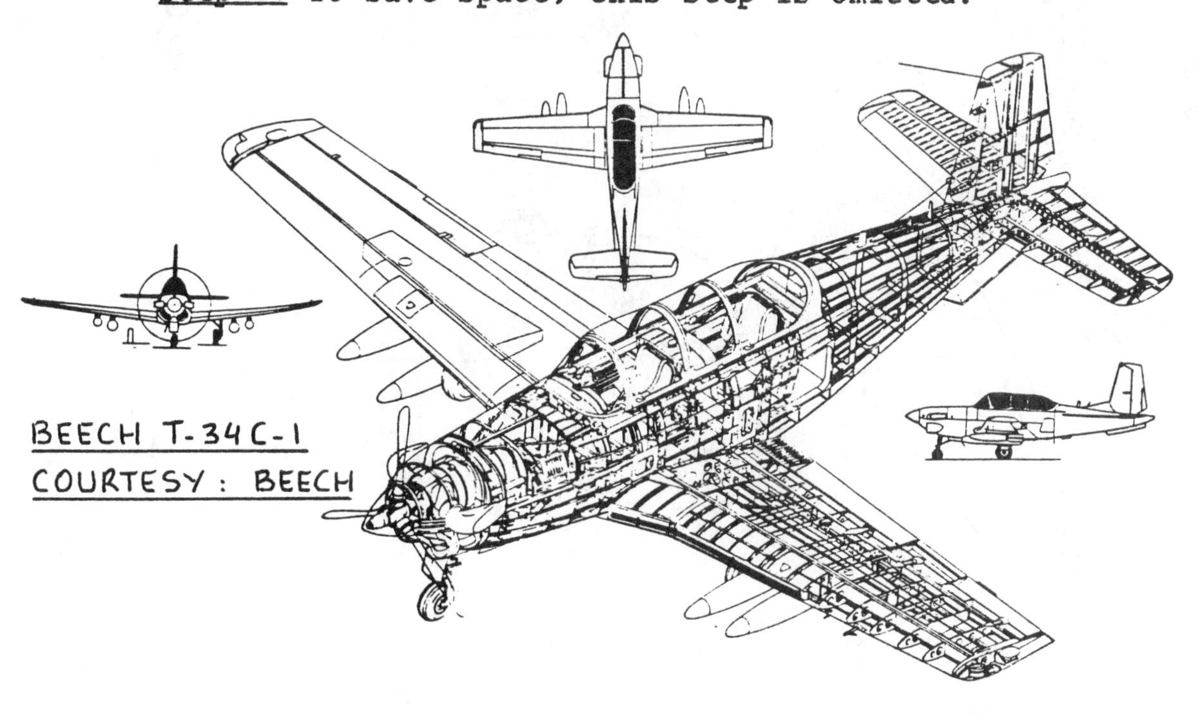

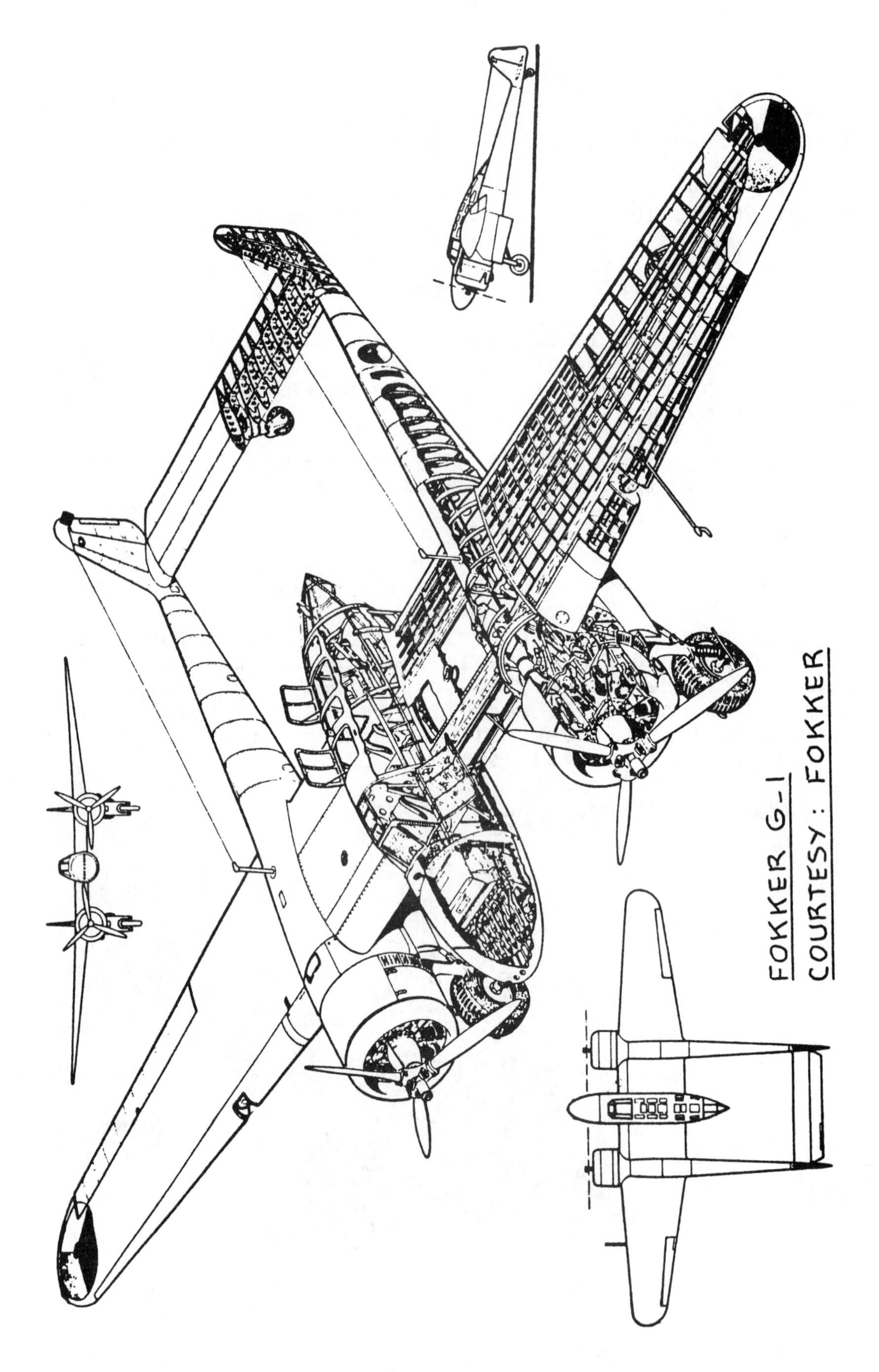
FOKKER G-1
COURTESY : FOKKER

3. CLASS I METHOD FOR ESTIMATING AIRPLANE INERTIAS

The purpose of this chapter is to provide a methodology for rapidly estimating airplane inertias. The emphasis is on rapid and on spending as few engineering manhours as possible. Methods which fit meet these objectives are referred to as Class I methods. They are used in conjunction with the first stage in the preliminary design process, the one referred to as 'p.d. sequence I' in Part II (Ref.2).

Section 3.1 presents a Class I method for estimating I_{xx}, I_{yy} and I_{zz}. These inertia moments are useful whenever it is necessary to evaluate undamped natural frequencies and/or motion time constants for airplanes during p.d. sequence I.

Example applications are discussed in Section 3.2.

3.1 ESTIMATING MOMENTS OF INERTIA WITH RADII OF GYRATION

The Class I method for airplane inertia estimation relies on the assumption, that within each airplane category it is possible to identify a radius of gyration, $R_{x,y,z}$ for the airplane. The moments of inertia of the airplane are then found from the following equations:

$$I_{xx} = (R_x)^2 W/g \quad (3.1)$$

$$I_{yy} = (R_y)^2 W/g \quad (3.2)$$

$$I_{zz} = (R_z)^2 W/g \quad (3.3)$$

Research in References 9, 10 and 11 has shown that a non-dimensional radius of gyration can be associated with each R component in the following manner:

$$\bar{R}_x = 2R_x/b \quad (3.4)$$

$$\bar{R}_y = 2R_y/L \quad (3.5)$$

$$\bar{R}_z = 2R_z/e, \text{ with: } e = (b + L)/2 \quad (3.6)$$

The quantities b and L in Eqns.(3.4) and (3.5) are the wing span and the overall airplane length respectively.

Airplanes of the same mission orientation tend to have similar values for the non-dimensional radius of gyration. Tables B.1 through B.12 (See Appendix B) present numerical values for these non-dimensional radii of gyration for different types of airplanes.

The procedure for estimating inertias therefore boils down to the following simple steps:

Step 1: List the values of W_{TO}, W_E, b, L and e for the airplane being designed.

Step 2: Identify which type of airplane in Tables B.1 through B.12 best 'fit' the airplane being designed.

Step 3: Select values for the non-dimensional radii of gyration corresponding to W_{TO} and W_E. It must be kept in mind that the distribution of the mass difference between W_{TO} and W_E is more important than the mass difference itself.

Acquiring the knowledge of what the airplanes in Tables B.1 through B.12 are like is therefore essential. As usual, Jane's (Ref. 8) is the source for acquiring that knowledge.

Step 4: Compute the airplane moments of inertia from:

$$I_{xx} = b^2 W(\bar{R}_x)^2/4g \tag{3.7}$$

$$I_{yy} = L^2 W(\bar{R}_y)^2/4g \tag{3.8}$$

$$I_{zz} = e^2 W(\bar{R}_z)^2/4g \tag{3.9}$$

Values for b and for L follow from the airplane threeview. The value for e follows from Eqn.(3.6).

The reader will have noted that there is no rapid method for evaluating I_{xz}. This product of inertia can be realistically evaluated only from a Class II weight and balance analysis. Such an analysis is presented in Chapter 9. In the first stages of preliminary design I_{xz} is not usually important. Therefore, it is normally ignored until later stages in the design process.

Step 5: Compare the estimated inertias of Step 4 with the data of Figures 3.1 through 3.3. If the comparison is poor, find an explanation and/or make adjustments.

Step 6: Document the results obtained in Steps 1 through 5 in a brief, descriptive report. Include illustrations where necessary.

3.2 EXAMPLE APPLICATIONS

Three example applications will now be discussed:

3.2.1 Twin Engine Propeller Driven Airplane: Selene
3.2.2 Jet Transport: Ourania
3.2.3 Fighter: Eris

3.2.1 Twin Engine Propeller Driven Airplane

Step 1: The following information is available for the Selene airplane:

W_{TO} = 7,900 lbs W_E = 4,900 lbs b = 37.1 ft

L = 43.0 ft e = 40.05 ft (Part II, p.247, p.297)

Step 2: From Table B3 (Appendix B) the following airplanes are judged to be comparable to the Selene in terms of mass distribution: Beech D18S, Cessna 404 and Cessna 441.

Step 3: From Table B3 (Appendix B) it is estimated that the following non-dimensional radii of gyration apply to the Selene:

$\bar{R}_x = 0.30$ $\bar{R}_y = 0.34$ $\bar{R}_z = 0.40$

Step 4: With Eqns.(3.7) through (3.9) the following moments of inertia can now be calculated:

At W_{TO}:

$I_{xx} = 37.1^2 x 7{,}900 x 0.30^2/4 x 32.2 = 7{,}598 \text{ slugft}^2$

$I_{yy} = 43.0^2 x 7{,}900 x 0.34^2/4 x 32.2 = 13{,}109 \text{ slugft}^2$

$I_{zz} = 40.05^2 x 7{,}900 x 0.40^2/4 x 32.2 = 15{,}741 \text{ slugft}^2$

At W_E:

$I_{xx} = (4,900/7,900)x7,598 = 4,713$ slugft^2

$I_{yy} = (4,900/7,900)x13,109 = 8,131$ slugft^2

$I_{zz} = (4,900/7,900)x15,741 = 9,763$ slugft^2

Step 5: Figures 3.1 through 3.3 show that the inertia estimates of Step 4 are reasonable.

Step 6: This step has been omitted to save space.

3.2.2 Jet Transport

Step 1: The following information is available for the Ourania airplane:

$W_{TO} = 127,000$ lbs $W_E = 68,450$ lbs $b = 113.8$ ft

$L = 127.0$ ft $e = 120.4$ ft   (Part II, p.251, p.299)

Step 2: From Table B7a (Appendix B) the following airplanes are judged to be comparable to the Ourania in terms of mass distribution: Convair 880, Convair 990, Boeing 737-200, McDonnell Douglas DC8.

Step 3: From Table B7a (Appendix B) it is estimated that the following non-dimensional radii of gyration apply to the Ourania:

At W_{TO}: $\bar{R}_x = 0.25$ $\bar{R}_y = 0.38$ $\bar{R}_z = 0.46$

At W_E: $\bar{R}_x = 0.27$ $\bar{R}_y = 0.46$ $\bar{R}_z = 0.52$

Step 4: With Eqns.(3.7) through (3.9) the following moments of inertia can now be calculated:

At W_{TO}:

$I_{xx} = 113.8^2 x127,000x0.25^2/4x32.2 = 798,090$ slugft^2

$I_{yy} = 127.0^2 x127,000x0.38^2/4x32.2 = 2,296,479$ slgft^2

$I_{zz} = 120.4^2 x127,000x0.46^2/4x32.2 = 3,024,520$ slgft^2

At W_E:

I_{xx} = $113.8^2 x 68,450 x 0.27^2 / 4 x 32.2$ = 501,730 $slugft^2$

I_{yy} = $127.0^2 x 68,450 x 0.46^2 / 4 x 32.2$ = 1,813,764 $slugft^2$

I_{zz} = $120.4^2 x 68,450 x 0.52^2 / 4 x 32.2$ = 2,083,134 $slugft^2$

Step 5: Comparison with Figures 3.1 through 3.3 indicates that the inertia estimates of Step 4 are reasonable.

Step 6: To save space, this step has been omitted.

3.2.3 Fighter

Step 1: The following information is available for the Eris airplane:

W_{TO} = 64,905 lbs W_E = 33,500 lbs   b = 68.7 ft

L = 50.7 ft   e = 59.7 ft   (Part II, p.255, p.301)

Step 2: From Table B9a (Appendix B) the following airplanes are judged to be comparable to the Eris in terms of mass distribution: DH Vampire 20 and Gloster Meteor II. The reader should note that the Vampire is the only jet fighter in Table B9a with a twin boom configuration.

Step 3: From Table B9a (Appendix B) it is estimated that the following non-dimensional radii of gyration apply to the Eris:

$\bar{R}_x$ = 0.29 $\bar{R}_y$ = 0.32 $\bar{R}_z$ = 0.40

Step 4: With Eqns.(3.7) through (3.9) the following moments of inertia can now be calculated:

At W_{TO}:

I_{xx} = $68.7^2 x 64,905 x 0.29^2 / 4 x 32.2$ = 200,019 $slugft^2$

I_{yy} = $50.7^2 x 64,905 x 0.32^2 / 4 x 32.2$ = 132,641 $slugft^2$

I_{zz} = $59.7^2 x 64,905 x 0.40^2 / 4 x 32.2$ = 287,363 $slugft^2$

At W_E:

$$I_{xx} = 68.7^2 x33,500x0.29^2/4x32.2 = 103,237 \text{ slugft}^2$$

$$I_{yy} = 50.7^2 x33,500x0.32^2/4x32.2 = 68,461 \text{ slugft}^2$$

$$I_{zz} = 59.7^2 x33,500x0.40^2/4x32.2 = 148,319 \text{ slugft}^2$$

Step 5: Comparison of the results of Step 4 with Figures 3.1 through 3.3 indicate that the inertia estimates are reasonable.

Step 6: This step has been omitted to save space.

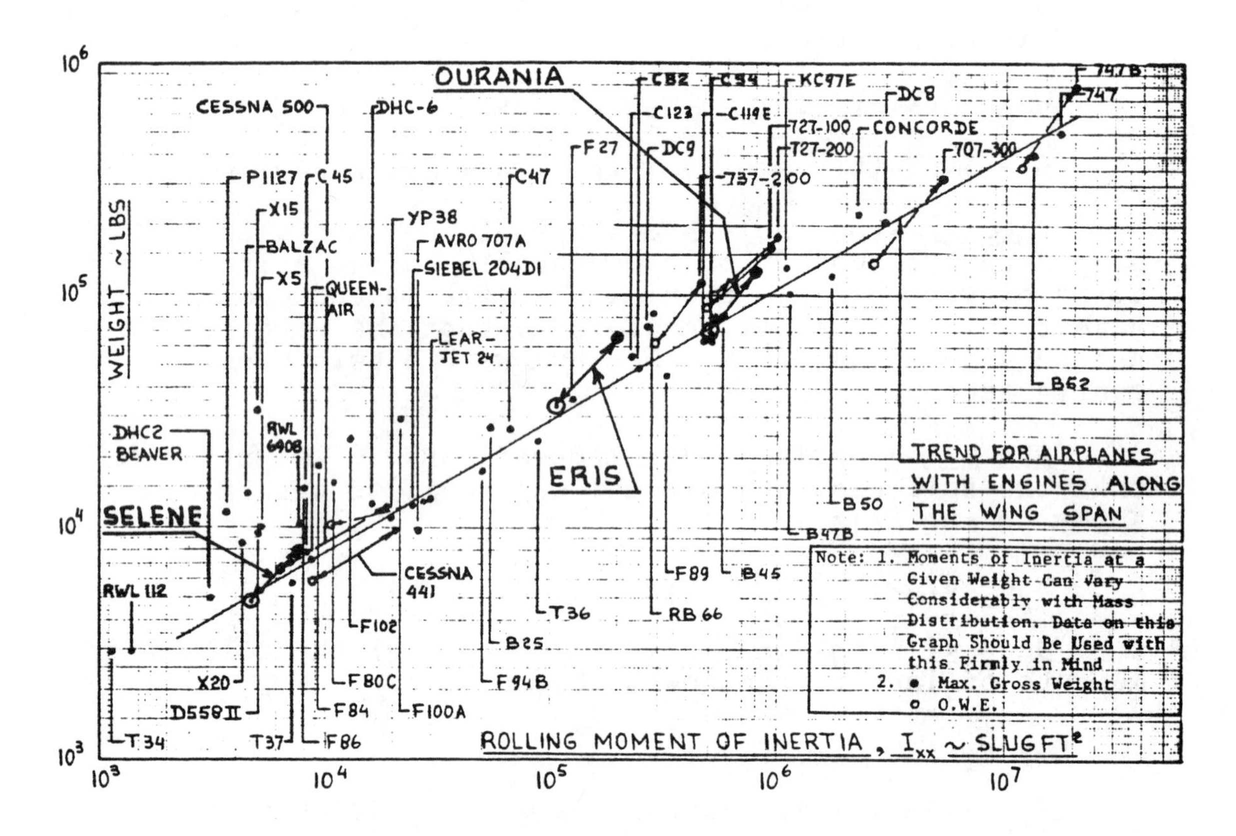

Figure 3.1 Correlation of Rolling Moments of Inertia with Weight

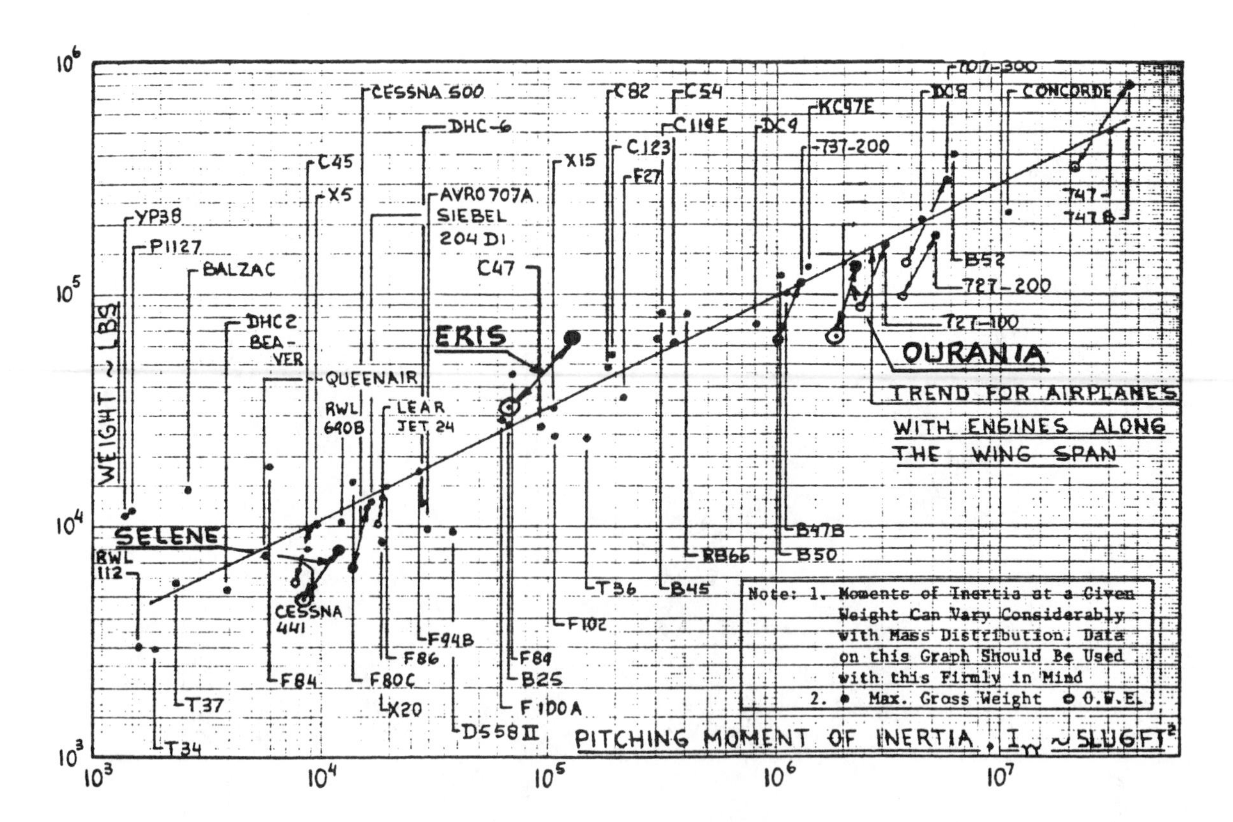

Figure 3.2 Correlation of Pitching Moments of Inertia with Weight

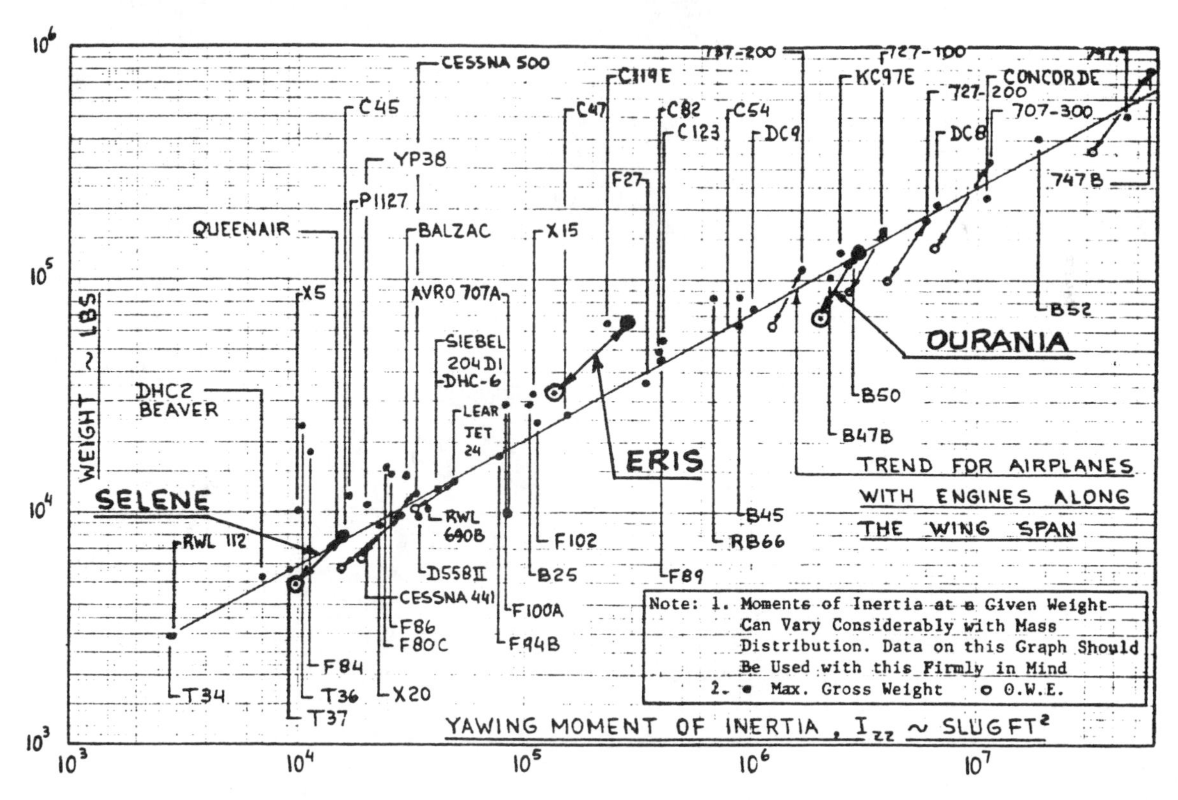

Figure 3.3 Correlation of Yawing Moments of Inertia with Weight

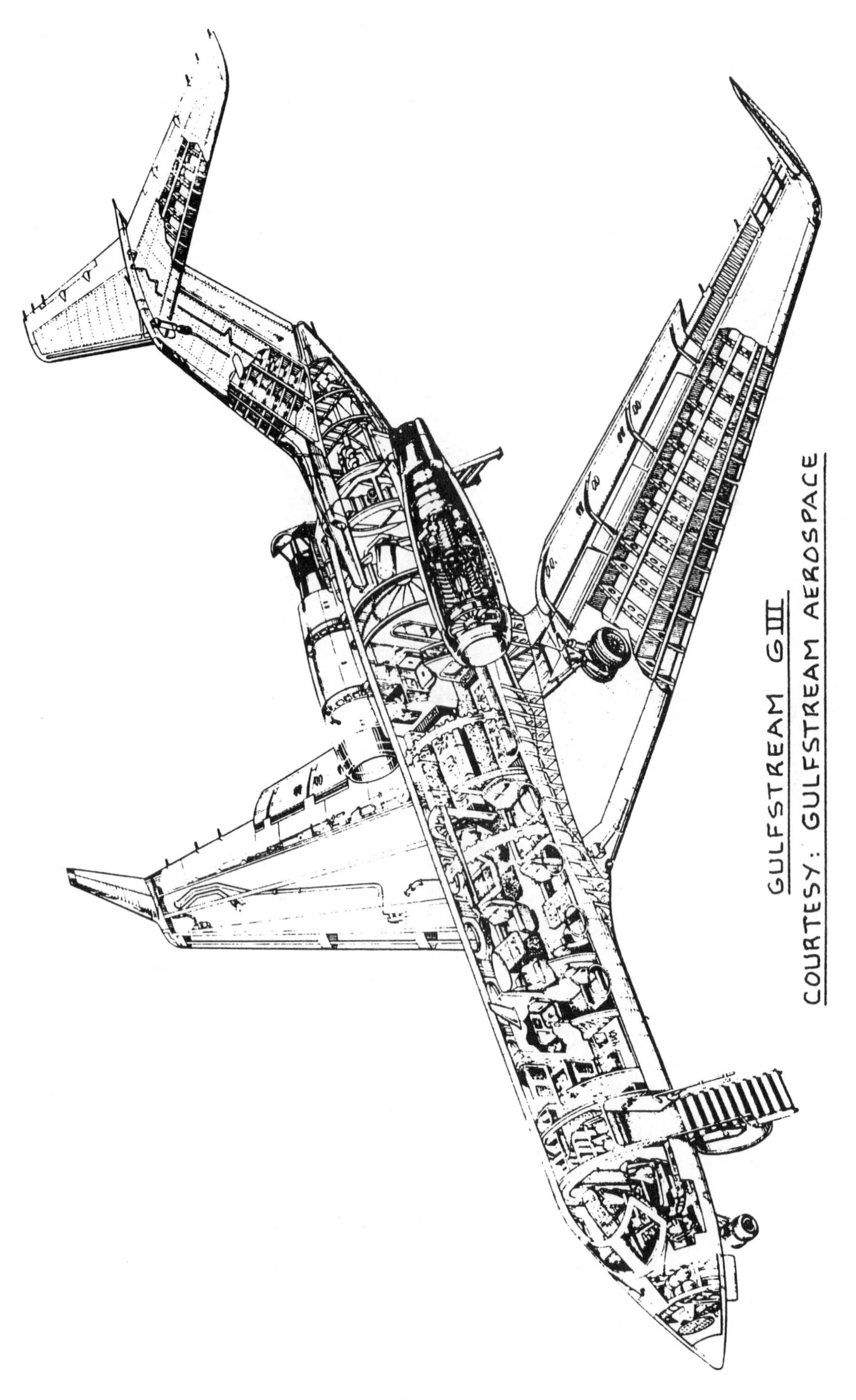

GULFSTREAM G III
COURTESY: GULFSTREAM AEROSPACE

4. CLASS II METHOD FOR ESTIMATING AIRPLANE COMPONENT WEIGHTS

The purpose of this chapter is to present a Class II method for estimating airplane component weights. Class II methods are those used in conjunction with preliminary design sequence II as defined in Part II, pp 18-23. The Class II weight estimating method accounts for such details as:

1. Airplane take-off gross weight

2. Wing and empennage design parameters such as:

   a. area
   b. sweep angle,
   c. taper ratio, λ
   d. thickness ratio, t/c

3. Load factor, n_{lim} or n_{ult}

4. Design cruise and/or dive speed, V_C or V_D

   Note: items 3 and 4 follow from a V-n diagram.

5. Fuselage configuration and interior requirements

6. Powerplant installation

7. Landing gear design and disposition

8. Systems requirements

9. Preliminary structural arrangement

To apply the Class II method for estimating component weights requires a fairly comprehensive knowledge about the airplane being designed. This knowledge was developed as a result of p.d. sequence I, discussed in Part II, pp 11-18.

Almost all airframe manufacturers have developed their own Class II methods for estimating airplane component weights. Many of these methods are proprietary. The Class II methods used in this text are based on those of References 12, 13 and 14. These methods employ empirical equations which relate component weights to airplane design characteristics such as items 1-9 above.

The following basic weight definition from Part I (Eqn.2.17) will be used:

$$W_{TO} = W_E + W_F + W_{PL} + W_{tfo} + W_{crew}, \tag{4.1}$$

where: W_E = empty weight, defined by Eqn.(4.2).

W_F = mission fuel weight, defined by: Eqn.(2.15) in Part I.

W_{PL} = payload weight, defined by the mission specification and on page 8, Part I.

W_{tfo} = weight of trapped fuel and oil, found from p.7, Part I.

W_{crew} = crew weight, defined by the mission specification and on page 8, Part I.

The Class II weight estimating method to be developed here will focus on estimating the components of empty weight, W_E which are defined as:

$$W_E = W_{struct} + W_{pwr} + W_{feq}, \tag{4.2}$$

where: W_{struct} = structure weight, discussed in Chapter 5.

W_{pwr} = powerplant weight, discussed in Chapter 6.

W_{feq} = fixed equipment weight, discussed in Chapter 7.

In Chapters 5-7 the specific Class II methods are identified as follows:

1. Cessna method: from Ref.12
2. USAF method from Ref.13
3. GD (General Dynamics) method from Ref.13
4. Torenbeek method from Ref.14

Section 4.1 presents a step-by-step procedure for using the Class II weight estimation method.

Section 4.2 presents a method for constructing the V-n diagram, needed in several of the weight equations employed in Chapters 5-7.

Example applications are presented in Section 4.3.

4.1 A METHOD FOR ESTIMATING AIRPLANE COMPONENT WEIGHTS WITH WEIGHT EQUATIONS

In this section a step-by-step procedure is presented for estimating airplane component weights and use these weights in estimating airplane empty weight, W_E.

As will be seen, this procedure is iterative. The reason is, that almost all airplane component weights themselves are a function of W_{TO}. A first estimate for W_{TO} was obtained during the preliminary sizing of the airplane. The reader will have noticed that during the Class I weight estimates (Chapter 2), the original estimate of W_{TO} remained unaltered. That will no longer be the case in the Class II method.

The method is presented as part of Step 21 in p.d. sequence II, as outlined on p.19 of Part II.

For the inexperienced reader, it is suggested that the following procedure be followed exactly as suggested.

<u>Step 1:</u> List all airplane components for which the weights are already known and tabulate their weights. This information can normally be obtained from the mission specification.

Typical items of known weight are:

1. Payload
2. Crew
3. Certain operational systems
4. Certain military loads
5. Engines (these are sometimes specified)

<u>Step 2:</u> List all airplane components for which the weights will have to be estimated. This list will contain at least the same items used in Class I. However, particularly in the systems area the list will contain much more detail at this point.

In preparing this list, use the groupings of components as indicated by Eqn.(4.2). Subdivision of these groupings should be done in accordance with Chapters 5-7, Eqns.(5.1), (6.1) and (7.1).

Step 3: Refer to the structural arrangement drawing prepared under Step 19, p.19, Part II.

The initial structural arrangement drawing is needed to identify those areas of the structure where special provisions were made or where, because of a clever structural arrangement a weight saving can be claimed.

Step 4: Determine from the tabulation below which weight estimation category best represents the airplane being designed.

Airplane Type	Weight Category for Component Weight Estimation Equations
1. Homebuilts	General Aviation Airplanes
2. Single Engine Props	General Aviation Airplanes
3. Twin Engine Props	General Aviation Airplanes
4. Agricultural	General Aviation Airplanes
5. Business Jets	Commercial Transports
6. Regional Turboprops	
below 12,500 lbs	General Aviation Airplanes
above 12,500 lbs	Commercial Transports
7. Jet Transports	Commercial Transports
8. Military Trainers	
low speed	General Aviation Airplanes
high speed	Fighter and Attack Airplanes
9. Fighters	Fighter and Attack Airplanes
10. Military Patrol, Bomb and Transport Airplanes	Military Patrol, Bomb and Transport Airplanes
11. Flying boats, Amphibious and Float Airplanes	
small and low speed	General Aviation Airplanes
large and high speed	Commercial Transports and/or Mil.Patr., Bomb and Transp.
12. Supersonic cruise	
Commercial	Commercial Transports, but use Fighter inlet data
Fighter and Attack	Fighter and Attack
Patrol, Bomb, Transp.	Mil.Patr., Bomb and Transp.

The weight estimation equations in Chapters 5-7 are all given in terms of the categories on the right side of the above table.

Step 5: Determine which equations in Chapters 5-7 apply to the airplane for which the Class II weight estimate is to be made. List these equations for each weight component.

Step 6: Make a list of all required input data needed in the equations of Step 5.

Step 7: Compute the component weights with the applicable equations of Step 5.

Notes:

1. The reader will observe that Chapters 5-7 often contain more than one equation to estimate the weight of a particular component. In that case estimate the weights with all applicable equations and use an average.

2. Sometimes it is desirable to 'calibrate' a component weight equation with the help of known weight data from existing airplanes. The component weight data of Appendix A can be used for this purpose. Calibration is done by applying the weight equations to the appropriate components and comparing the answers with the actual weight data of Appendix A. The so-called 'fudge-constants' which appear in all Class II weight equations can then be altered to obtain a better estimate. The reader should be careful and only use this 'calibration' method in conjunction with airplanes which have similar missions.

3. In the systems area, there are not enough reliable equations available. In that case it is desirable to also estimate the average applicable weight fraction for each system component. This can be done with the data in Appendix A. The examples in Section 4.3 show how this is done.

Step 8: Add all component weights and obtain an estimate for W_E, from Eqn.(4.2).

Step 9: Compute a new value for W_{TO} with Eqn.(4.1),

but: 1. use for W_E the value obtained in Step 8.

2. use for W_F a value obtained from

Eqn.(2.15) in Part I. This implies that the mission fuel needed must be adjusted for the new value of W_{TO}. The result is:

$W_{TO} =$ (4.3)

$$= (W_E + W_{PL} + W_{crew})/\{M_{ff}(1 + M_{res}) - M_{res} - M_{tfo}\}$$

Values for M_{ff}, M_{res} and M_{tfo} were already obtained during the preliminary sizing work described in Chapter 2 of Part I. These fractions may have changed if, during the Class I drag polar analysis of Chapter 12, Part II a significant change in L/D was discovered. In that case it was recommended in Step 14, Part II (p.16-17) to redo the preliminary sizing. This in turn would result in new values for the fractions in Eqn.(4.3).

Step 10: Use this new estimate for W_{TO} to iterate back through Steps 7-9 until the W_{TO} values agree within 0.5 percent.

Notes:

1. If the new value of W_{TO} obtained in Step 9 differs from the original one by more than 5 percent it will be necessary to account for the effect on required engine thrust (or power) at take-off. This in turn will affect the engine weight.

2. Accounting for a change in required take-off thrust (or power) may be done by using the ratio $(T/W)_{TO}$ (or $(W/P)_{TO}$ obtained from the preliminary sizing process of Chapter 3, Part I.

Step 11: Document all calculations including all assumptions made, all decisions made and all interpretations made in a brief, descriptive report. Where needed, include clearly drawn sketches.

Include a final Class II weight statement, using the groupings suggested by Eqns.(4.2), (5.1), (6.1) and (7.1).

4.2 METHODS FOR CONSTRUCTING V-n DIAGRAMS

In this section a step-by-step procedure is presented for constructing V-n diagrams for the following types of airplanes:

4.2.1 FAR 23 Certified Airplanes
4.2.2 FAR 25 Certified Airplanes
4.2.3 Military Airplanes

Example applications for three airplanes are provided in sub-section 4.2.4.

The V-n diagrams are used to determine design limit and design ultimate load factors as well as the corresponding speeds to which airplane structures are designed. As will be seen in the Class II weight equations of Chapters 5-7, many require as input a design load factor and/or a design speed.

Important notes:

1. The V-n diagrams given here are simplified versions of those defined in Refs 15-17. They should be used only in conjunction with Class II weight estimation methods.

2. In the Class II method only flaps-up cases are considered.

4.2.1 V-n Diagram for FAR 23 Certified Airplanes

Reference 15, in Part 23.335 presents the V-n diagram shown in Figure 4.1. The following definitions apply to the various speeds given in the diagram:

Note: all speeds are normally given in KEAS.

V_S = +1g stall speed or the minimum speed at which the airplane is controllable

V_C = design cruising speed

V_D = design diving speed

V_A = design maneuvering speed

Determination of these speeds and determination of the critical points A, C, D, E, F and G in Figure 4.1 is discussed in sub-sub-sections 4.2.1.1 through 4.2.1.7.

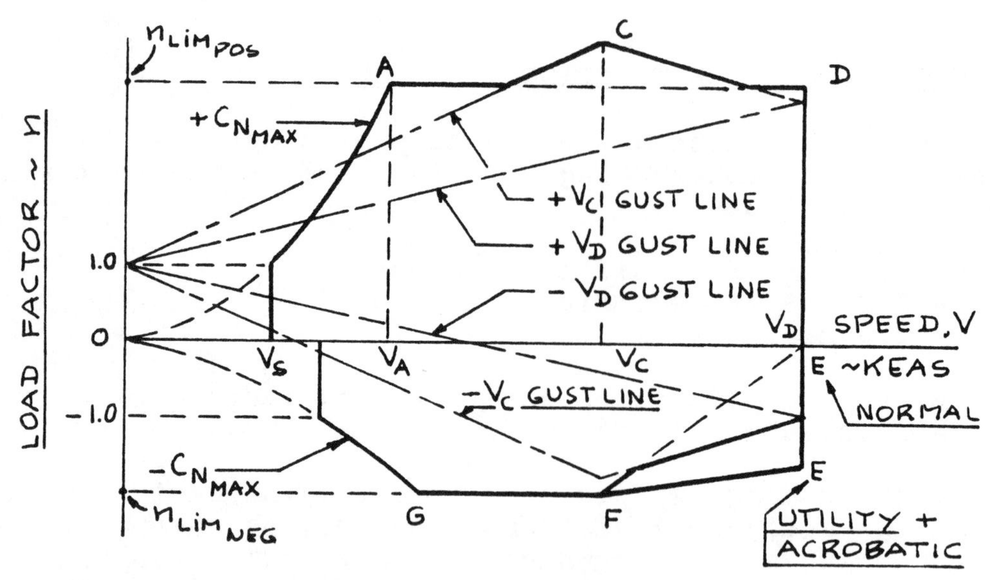

Figure 4.1 V-n Diagram According to FAR 23

4.2.1.1 Determination of +1g stall speed, V_S

$$V_S = \{2(GW/S)/\rho C_{N_{max}}\}^{1/2}, \tag{4.4}$$

where: GW = flight design gross weight in lbs

S = wing area in ft^2

ρ = air density in $slugs/ft^3$

$C_{N_{max}}$ = maximum normal force coefficient.

The maximum normal force coefficient follows from:

$$C_{N_{max}} = \{(C_{L_{max}})^2 + (C_{D_{at\ C_{L_{max}}}})^2\}^{1/2} \tag{4.5}$$

In preliminary design it is acceptable to set:

$$C_{N_{max}} = 1.1 C_{L_{max}} \tag{4.6}$$

4.2.1.2 Determination of design cruising speed, V_C

$$V_C \geqslant k_c (GW/S)^{1/2}, \tag{4.7}$$

where the constant k_c takes on the following values:

k_c = 33 for normal and utility category airplanes up to W/S = 20 psf.

k_C varies linearly from 33 to 28.6 as W/S varies from 20 to 100, for normal and utility category airplanes.

$k_C = 36$ for acrobatic category airplanes.

Note: V_C need not be more than $0.9V_H$, where V_H is the maximum level speed obtained with maximum power or with maximum thrust.

4.2.1.3 Determination of design diving speed, V_D

$$V_D \text{ (or } M_D) \geqslant 1.25V_C \text{ (or } 1.25M_C), \tag{4.8}$$

where: V_C follows from Eqn.(4.7).

4.2.1.4 Determination of design maneuvering speed, V_A

$$V_A \geqslant V_S n_{lim}^{1/2}, \tag{4.9}$$

where: n_{lim} is the limit maneuvering load factor given by Eqn.(4.13).

Note: V_A need not exceed V_C

4.2.1.5 Determination of negative stall speed line

$$V_{S_{neg}} = \{2(GW/S)/\rho C_{N_{max_{neg}}}\}^{1/2}, \tag{4.10}$$

where $C_{N_{max_{neg}}}$ is given by:

$$C_{N_{max_{neg}}} = \{(C_{L_{max_{neg}}})^2 + (C_{D_{atC_{L_{max_{neg}}}}})^2\}^{1/2} \tag{4.11}$$

In preliminary design it is acceptable to use:

$$C_{N_{max_{neg}}} = 1.1C_{L_{max_{neg}}}, \tag{4.12}$$

where: $C_{L_{max_{neg}}}$ is the maximum negative lift coefficient.

4.2.1.6 Determination of design limit load factor, n_{lim}

The positive, design limit load factor is given by:

$$n_{lim_{pos}} \geqslant 2.1 + 24{,}000/(GW + 10{,}000) \quad (4.13)$$

Exceptions:

n_{lim} need not be greater than 3.8

n_{lim} = 4.4 for utility category airplanes

n_{lim} = 6.0 for acrobatic airplanes

The negative, design limit load factor is given by:

$$n_{lim_{neg}} \geqslant 0.4n_{lim} \text{ for normal and for utility category airplanes} \quad (4.14)$$

$$\geqslant 0.5n_{lim} \text{ for acrobatic airplanes} \quad (4.15)$$

4.2.1.7 Construction of gust load factor lines in Fig.4.1

The gust load factor lines in Figure 4.1 are defined by the following equation:

$$n_{lim} = 1 + (K_g U_{de} V C_{L_\alpha})/498(GW/S), \quad (4.16)$$

where: K_g is the gust alleviation factor given by:

$$K_g = 0.88\mu_g/(5.3 + \mu_g), \quad (4.17)$$

where:

$$\mu_g = 2(GW/S)/\rho \bar{c} g C_{L_\alpha} \quad (4.18)$$

The derived gust velocity, U_{de} is defined as follows:

For the V_C gust lines:

U_{de} = 50 fps between sealevel and 20,000 ft

U_{de} = 66.67 - 0.000833h between 20,000 and 50,000 ft

For the V_D gust lines:

U_{de} = 25 fps between sealevel and 20,000 ft

U_{de} = 33.34 - 0.000417h between 20,000 and 50,000 ft

4.2.2 V-n Diagram for FAR 25 Certified Airplanes

Reference 16, in Part 25.335 presents the two V-n diagrams shown in Figures 4.2a and 4.2b. The following definitions apply to the various speeds given in the diagrams:

Note: all speeds are normally given in KEAS.

V_{S_1} = +1-g stall speed or the minimum steady flight speed which can be obtained

V_C = design cruising speed

V_D = design diving speed

V_A = design maneuvering speed

V_B = design speed for maximum gust intensity

Determination of these speeds and determination of the critical points A, D, E, F, H, B', C', D', E', F' and G' is discussed in sub-sub-sections 4.2.2.1 through 4.2.2.8.

4.2.2.1 Determination of +1g stall speed, V_{S_1}

$$V_{S_1} = \{2(GW/S)/\rho C_{N_{max}}\}^{1/2}, \qquad (4.19)$$

where: GW = flight design gross weight in lbs

S = wing area in ft^2

ρ = air density in $slugs/ft^3$

$C_{N_{max}}$ = maximum normal force coefficient, as computed from Eqn.(4.5) or (4.6).

4.2.2.2 Determination of design cruising speed, V_C

V_C must be sufficiently greater than V_B to provide for inadvertent speed increases likely to occur as a result of severe atmospheric turbulence. For V_B, see sub-sub-section 4.2.2.5.

$$V_C \geqslant V_B + 43 \text{ kts} \qquad (4.20)$$

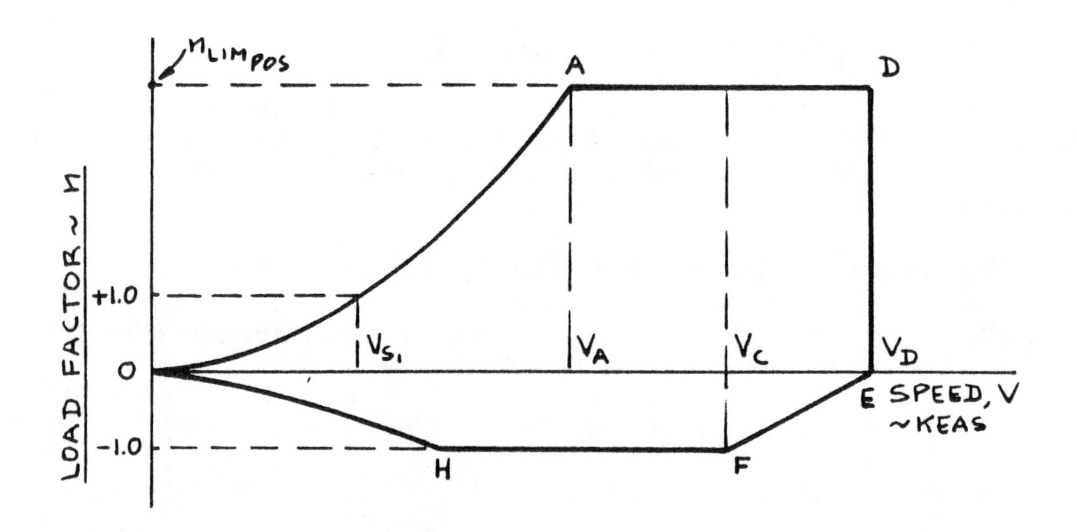

Figure 4.2a V-n Maneuver Diagram According to FAR 25

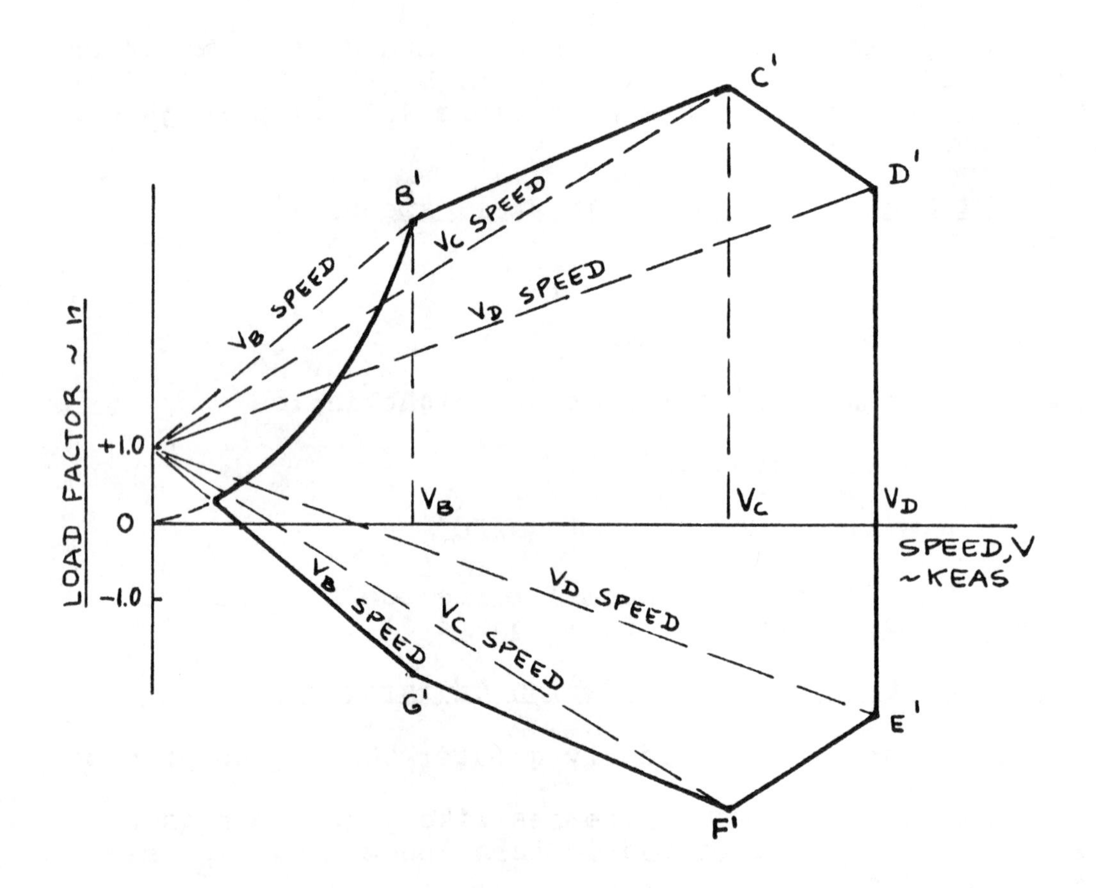

Figure 4.2b V-n Gust Diagram According to FAR25

4.2.2.3 Determination of design diving speed, V_D

$$V_D \text{ (or } M_D) \geq 1.25V_C \text{ (or } 1.25M_C) \quad (4.21)$$

where: V_C follows from Eqn.(4.20).

4.2.2.4 Determination of design maneuvering speed, V_A

$$V_A \geq V_{S_1} n_{lim}^{1/2}, \quad (4.22)$$

where: n_{lim} is the limit maneuvering load factor at V_C.

The limit maneuvering load factor in Eqn.(4.22) follows from 4.2.2.7 or from 4.2.2.8 depending on which is the more critical.

Note: V_A need not exceed V_C.

4.2.2.5 Determination of design speed for maximum gust intensity, V_B

V_B need not be greater than V_C.

V_B may not be less than the speed determined from the intersection of the $C_{N_{max}}$ line and the gustline marked V_B.

4.2.2.6 Determination of negative stall speed line

The negative stall speed line in Figure 4.2a is determined with the method of sub-sub-section 4.2.1.5.

4.2.2.7 Determination of design limit load factor, n_{lim}

The positive limit maneuvering load factor, $n_{lim_{pos}}$ is determined from:

$$n_{lim_{pos}} \geq 2.1 + \{24{,}000/(W + 10{,}000)\} \quad (4.23)$$

Exceptions:

$n_{lim_{pos}} \geq 2.5$ at all times

$n_{lim_{pos}}$ need not be greater than 3.8 at W_{TO}

The negative, design limit load factor is determined from:

$n_{lim_{neg}} \geq -1.0$ up to V_C

$n_{lim_{neg}}$ varies linearly from the value at V_C to zero at V_D

4.2.2.8 Construction of gust load factor lines in Fig.4.2b

The gust load factor lines in Figure 4.2b are arrived at with the help of Eqns.(4.16) through (4.18). The derived gust velocities, U_{de} in FAR 25 are as follows:

For the gust line marked V_B:

U_{de} = 66 fps between sealevel and 20,000 ft

U_{de} = 84.67 - 0.000933h between 20,000 and 50,000 ft

For the gust line marked V_C:

U_{de} = 50 fps between sealevel and 20,000 ft

U_{de} = 66.67 - 0.000833h between 20,000 and 50,000 ft

For the gust line marked V_D:

U_{de} = 25 fps between sealevel and 20,000 ft

U_{de} = 33.34 - 0.000417h between 20,000 and 50,000 ft

4.2.3 V-n Diagram for Military Airplanes

Reference 17, provides the V-n diagram given in Figure 4.3. The indicated limit load factors must not be less than those defined in Table 4.1.

The speeds in Figure 4.3 are normally given in KEAS and are defined as follows:

V_H = maximum level speed which can be attained at the combination of weight and altitude under consideration

V_L = maximum dive speed, typically $1.25V_H$

Gust lines are as in FAR 25. Gust induced load factors are normally not critical for military airplanes with limit load factors above 3.00.

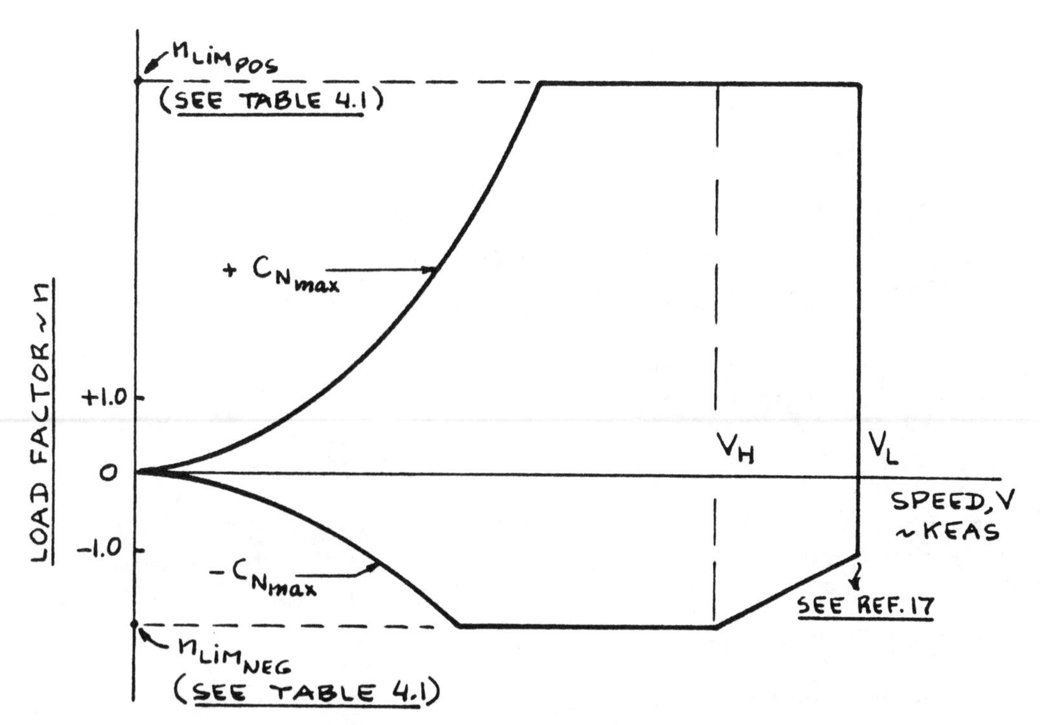

Figure 4.3 V-n Diagram According to MIL-A 8861(ASG)

Table 4.1 Limit Load Factors for Military Airplanes

Airplane Type		Limit Load Factor, n_{lim} at Flight Design Gross Weight, GW	
USAF	USN	Positive	Negative
Fighter		8.67	-3.00
Attack	Fighter, Attack, Trainer	7.33	-3.00
	Observation	6.00	-3.00
Trainer		5.67	-2.33
Utility	Utility	4.00	-2.00
Small Bomber		3.67	-1.67
Medium Bomber, Assault Transp.	Patrol, Weather, Anti-submarine, Reconnaissance	3.00	-1.00
Medium Transp.		2.50	-1.00
Heavy Bomber, Heavy Transp.		2.00	-1.00

4.2.4 Example Applications

The following example applications will be discussed:

4.2.4.1 Twin Engine Propeller Driven Airplane: Selene
4.2.4.2 Jet Transport: Ourania
4.2.4.3 Fighter:Eris

4.2.4.1 Twin Engine Propeller Driven Airplane

According to the mission specification (Table 2.17, Part I) this is a FAR 23 airplane. It will be assumed that under FAR 23 it will be certified under the normal category.

Determination of V_S:

Since $C_{L_{max}} = 1.7$ (Part I, p.178), it follows from Eqn.(4.6) that: $C_{N_{max}} = 1.1x1.7 = 1.87$.

Since $(W/S)_{TO} = 46$ psf (Part I, p.178), the value for stall speed as found from Eqn.(4.4) is:

$V_S = \{2x46/0.002378x1.87\}^{1/2} = 144$ fps $= 85$ kts.

Determination of V_C

The design wing loading for the Selene is 46 psf. This yields $k_c = 31.6$. With Eqn.(4.7) this in turn gives $V_C \geqslant 214$ kts.

The Selene was to have a cruise speed of 250 kts at 75 percent power at 10,000 ft (Part I, Table 2.17). For 100 percent power this would yield a maximum cruise speed which is a factor $(100/75)^{1/3} = 1.1$ higher, or 275 kts. According to sub-sub-section 4.2.1.2, V_C need not be higher than $0.9x275 = 248$ kts.

Thus: $V_C = 248$ kts.

Determination of V_D

According to sub-sub-section 4.2.1.3, the design dive speed is: $V_D = 1.25x248 = 310$ kts.

Determination of n_{lim}

The positive limit load factor of the Selene as given by Eqn.(4.13) is:

$n_{lim_{pos}} \geqslant 2.1 + \{24,000/(7,900 + 10,000)\} = 3.44$

The negative limit load factor as given by Eqn.(4.14) is:

$n_{lim_{neg}} = 0.4x3.44 = 1.38$

Determination of Gust Load Factor Lines, V_C and V_D

The overall airplane liftcurve slope, $C_{L_{\alpha}}$ can be shown to be 0.095 deg^{-1}= 5.44 rad^{-1}. With $\bar{c}$ = 4.92 ft (Table 13.1, Part II), the value of μ_g is: 44.8, according to Eqn.(4.18).

The gust alleviation factor follows from Eqn.(4.17) as:

$K_g = 0.88x44.8/(5.3 + 44.8) = 0.787$

The gust load factor lines now follow from Eqn.(4.16) as:

$n_{lim_{gust}} = 1 + 0.0094V$ for the V_C line and:

$n_{lim_{gust}} = 1 + 0.0047V$ for the V_D line.

Determination of V_A

From Eqn.(4.9): V_A $85x(3.44)^{1/2} = 158$ kts.

Determination of Negative Stall Line

It will be assumed that $C_{L_{max_{neg}}} = -1.18$. This yields $C_{N_{max_{neg}}} = -1.3$. Using Eqn.4.4 it is found that the negative 1g stall speed is 102 kts.

With these data it is now possible to draw the V-n diagram for the Selene. The result is shown in Fig.4.4.

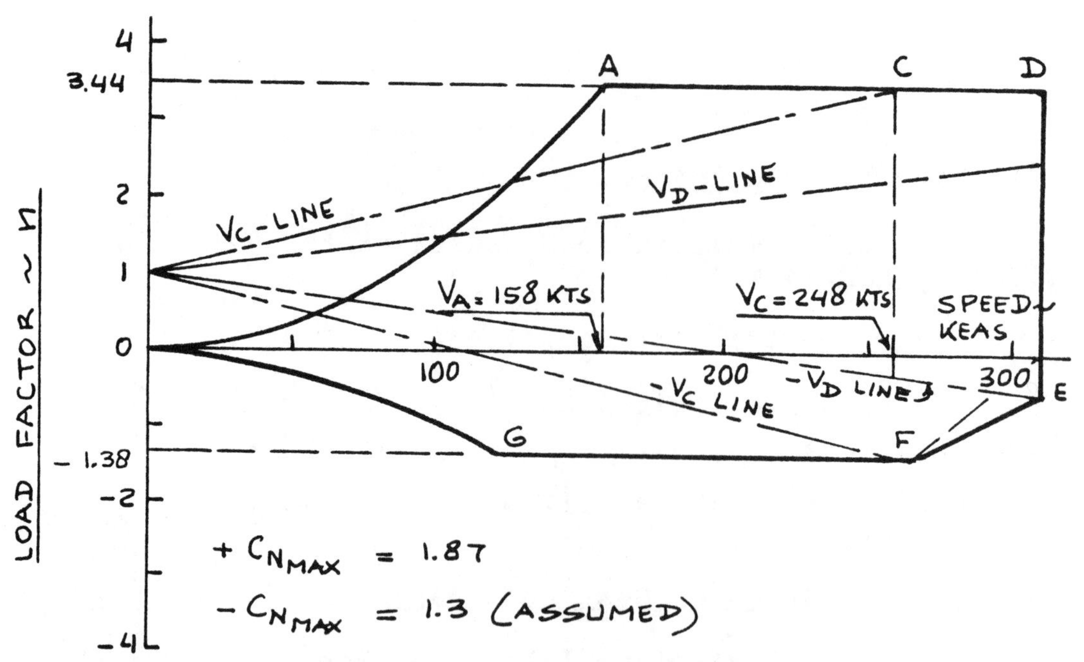

Figure 4.4 Example V-n Diagram for the Selene

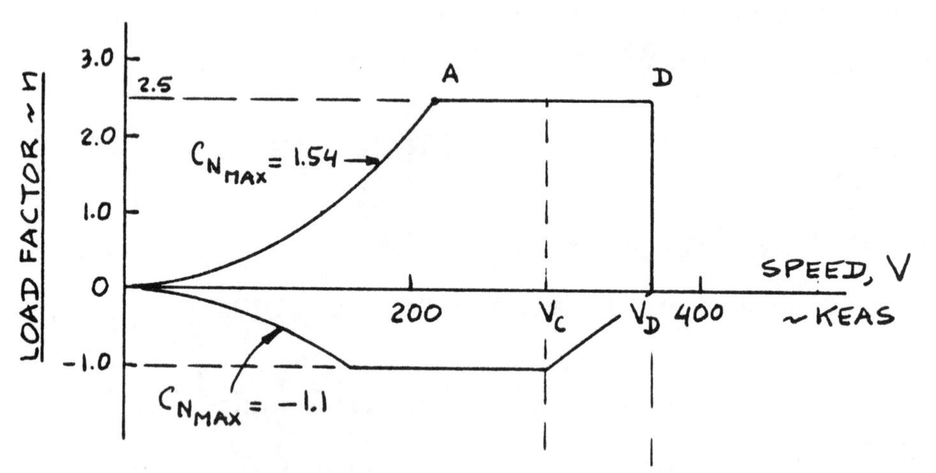

Figure 4.5a Example V-n Maneuver Diagram for the Ourania

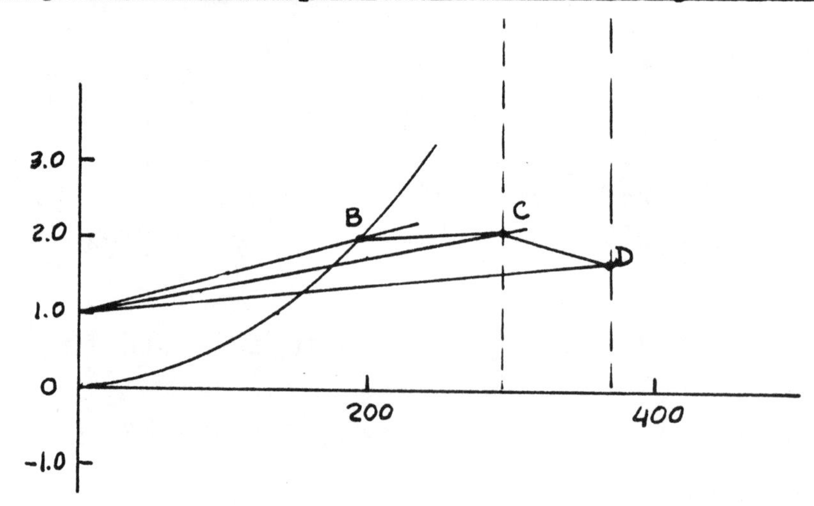

Figure 4.5b Example V-n Gust Diagram for the Ourania

4.2.4.2 Jet Transport

According to the mission specification (Table 2.18, Part I) this is a FAR 25 airplane.

Determination of V_{S_1} :

Since $C_{L_{max}}$ = 1.4 (Part I, p.184), it follows from Eqn.(4.6) that: $C_{N_{max}} = 1.1x1.4 = 1.54$.

Since $(W/S)_{TO}$ = 98 psf (Part I, p.184), the value for stall speed as found from Eqn.(4.4) is:

$V_S = \{2x98/0.002378x1.54\}^{1/2}$ = 231 fps = 137 kts.

Determination of V_A

V_A follows from the intersection of the +1g stall line and the +2.50 load factor line: V_A = 217 kts.

Determination of V_B

V_B follows from the intersect of the +1g stall line and the V_B gust line. This intersect will be determined upon calculation of the V_B gust line.

Determination of V_C

According to Eqn.(4.20): V_C V_B + 43 kts.

Therefore: V_C 195 + 43 = 238 kts. However, the mission specification of Table 2.18 (Part I) calls for a cruise speed of M = 0.82 at 35,000 ft. This corresponds to 483 kts at 35,000 ft or a dynamic pressure of 296 psf. At sealevel, the corresponding value in KEAS is 295 kts. Since this is larger 238 kts, V_C = 295 kts.

Determination of V_D

According to sub-sub-section 4.2.2.3, the design dive speed is: $V_D = 1.25xV_C = 1.25x295$ = 369 kts.

Determination of n_{lim}

The positive limit load factor of the Ourania as

given by Eqn.(4.23) is:

$n_{lim_{pos}}$ 2.1 +{24,000/(127,000 + 10,000)} = 2.28

The exceptions in sub-sub-section 4.2.2.7 demand that this load factor never be less than 2.5. Therefore:

$n_{lim_{pos}} = 2.5$

The negative limit load factor is -1 up to V_C and varies linearly to zero at V_D.

Determination of Gust Load Factor Lines, V_B, V_C and V_D

The overall airplane liftcurve slope, C_{L_α} can be shown to be 0.085 deg^{-1}= 4.87 rad^{-1}. With $\bar{c}$ = 12.5 ft (Table 13.2, Part II), the value of μ_g is: 42.0, according to Eqn.(4.18).

The gust alleviation factor follows from Eqn.(4.17) as:

$K_g = 0.88x42.0/(5.3 + 42.0) = 0.781$

The gust load factor lines now follow from Eqn.(4.16) as:

$n_{lim_{gust}} = 1 + 0.0051V$ for the V_B line,

$n_{lim_{gust}} = 1 + 0.0039V$ for the V_C line and:

$n_{lim_{gust}} = 1 + 0.0019V$ for the V_D line.

Determination of V_B

From the intersection of the +1g stall line with the V_B gust line it follows that V_B = 195 kts.

Determination of Negative Stall Line

It will be assumed that $C_{L_{max_{neg}}} = - 1.00$. This yields $C_{N_{max_{neg}}} = - 1.1$. Using Eqn.4.4 it is found that the negative 1g stall speed is 162 kts.

With these data it is now possible to draw the V-n diagram for the Ourania. The result is shown in Figures 4.5a and 4.5b.

4.2.4.3 Fighter

According to the mission specification (Table 2.19, Part I) the Eris is an attack fighter. From Table 4.1 it follows that $n_{lim_{pos}} = 7.33$ and $n_{lim_{neg}} = -3.0$.

The maximum level speed at sealevel is $V_H = 450$ kts.

The design dive speed, $V_L = 1.25 \times 450 = 563$ kts.

The gust lines are far within the maneuvering V-n diagram and are not computed for this airplane. Fig.4.6 presents the V-n diagram for the Eris.

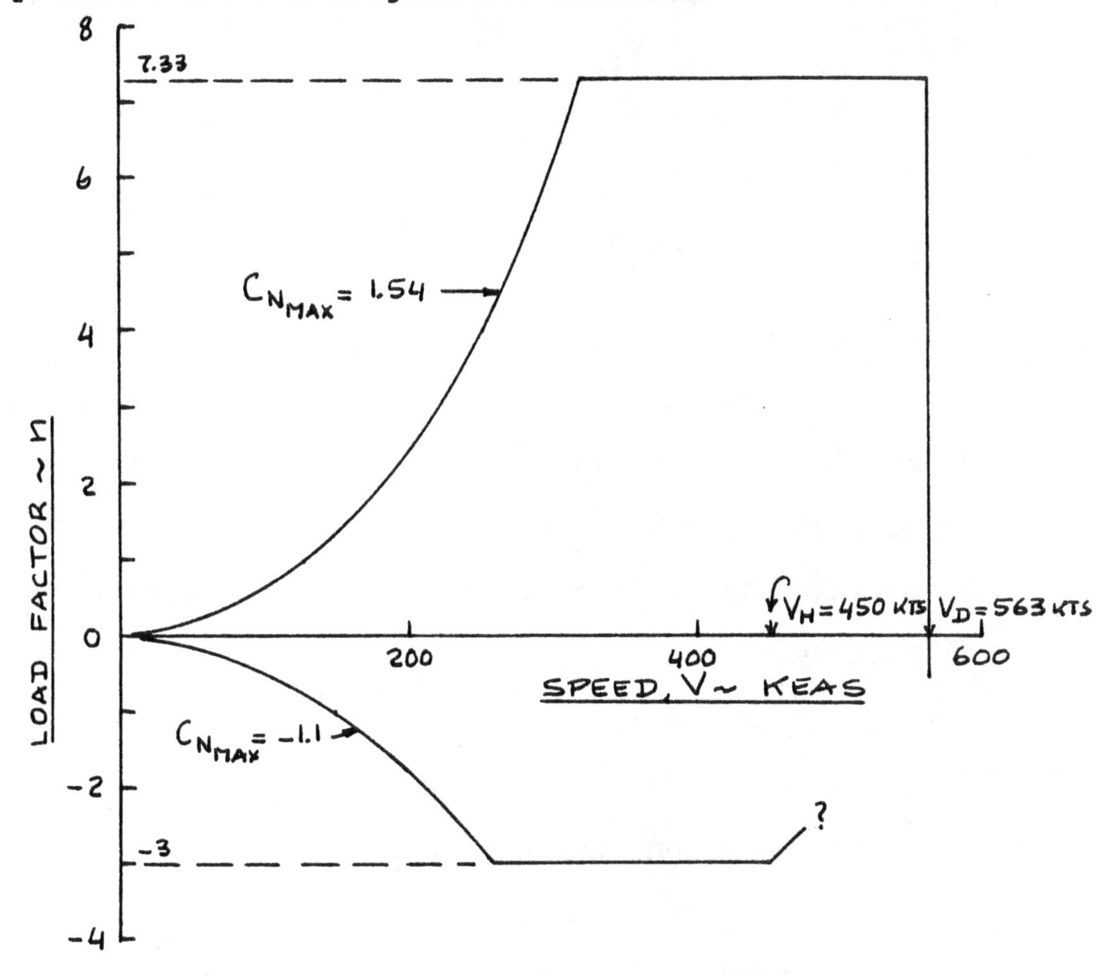

Figure 4.6 Example V-n Diagram for the Eris

4.3 EXAMPLE APPLICATIONS FOR CLASS II WEIGHT ESTIMATES

In this section, three example applications of the Class II weight estimation method described in Section 4.1 are discussed:

4.3.1 Twin Engine Propeller Driven Airplane: Selene
4.3.2 Jet Transport: Ourania
4.3.3 Fighter: Eris

4.3.1 Twin Engine Propeller Driven Airplane

Step 1: The following weight items are already known:

From Table 10.4, Part II:

Payload: $W_{PL} = 1,250$ lbs

Fuel: $W_F = 1,706$ lbs

TFO: $W_{tfo} = 44$ lbs

From Part II, p.135:

Engine dry weight: $W_e = 1,400$ lbs

Step 2: Weights need to be estimated for the following items:

Structural Weight, W_{struct}:

1) Wing 2) Adjustment for Fowler flaps 3) Empennage

4) Fuselage 5) Nacelles 6) Landing Gear

Powerplant Weight, W_{pwr}:

1) Engines 2) Air induction system 3) Propellers

4) Fuel System 5) Propulsion installation

Fixed Equipment Weight, W_{feq}:

1) Flight controls 2) Electrical system

3) Instrumentation, avionics and electronics

4) Air-conditioning and de-icing 5) Oxygen

6) Furnishings 7) Paint

Step 3: The structural arrangement drawing for the Selene is presented in Chapter 8 of Part III.

Step 4: From a weight estimating viewpoint this airplane falls in the General Aviation Airplane category.

Step 5: The following weight equations apply to the Selene:

W_{struct}: 1) Wing: Eqns (5.4) and (5.5)

2) Adjustment for Fowler flaps: an extra factor of 2 percent will be added in accordance with 5.2.2.2.

3) Empennage: Eqns (5.14) - (5.16)

4) Fuselage: Eqns (5.25) and (5.27)

5) Nacelles: Eqn. (5.33)

6) Landing Gear: Eqns (5.40) and (5.42)

W_{pwr}: 1) Engines: see Step 1.

2) Air induction system: Eqn.(6.8)

3) Propellers: Eqns (6.13) and (6.14)

4) Fuel System: Eqns (6.17) and (6.18)

5) Propulsion system: Eqns (6.3) and (6.4)

W_{feq}: 1) Flight control system: Eqns (7.1), (7.2) and (7.4).
Note: hydraulics and pneumatics are included in item 1).

2) Electrical system: Eqns (7.12) - (7.14)

3) Instrumentation, avionics and electronics: Eqn.(7.21)

4) Air-conditioning + de-icing: Eqn.(7.28)

5) Oxygen system: Eqn.(7.35)

6) Furnishings: Eqns (7.41) and (7.43)

7) Paint: Table A3.2a

Step 6: The following list itemizes all required input data for estimating the weight items listed in steps 2 and 5.

W_{TO} 7,900 lbs	n_{lim} = 3.44	S = 172 ft^2
V_C = 248 kts	V_D = 310 kts	n_{ult} = 5.16
A = 8	λ = 0.4	$\Lambda_{1/4}$ = 0 deg.
$(t/c)_m$ = 0.17	b = 37.1 ft	t_r = 1.13 ft
S_h = 58 ft^2	b_h = 14.9 ft	t_{r_h} = 0.53 ft
l_h = 24.3 ft		
S_v = 38 ft^2	b_v = 6.16 ft	t_{r_v} = 0.66 ft
l_f = 39.3 ft	w_f = 4.5 ft	h_f = 5.5 ft
K_f = 1.08	P_{TO} = 850 hp	l_{s_m} = 6.00 ft
W_L = 7,505 lbs	$n_{ult.l.}$ = 4.0	
K_{propl} = 31.92	N_p = 2	N_{bl} = 3
D_p = 7.8 ft		

Notes: 1) The value for n_{lim} follows from the V-n diagram of Figure 4.4.

2) Most data were obtained from Selene data listed in Part II. The reader is reminded that a detailed geometric definition may be found in Part II as Table 13.1, a Class I weight statement as Table 10.4. Detailed definitions of fuselage, wing, tails, landing gear and powerplant may be found in Chapters 4,5,6,7,8 and 9 respectively in Part II.

Step 7: Table 4.1 lists all weights computed as part of the Class II weight estimation process. Observe that the Class I weight estimates (computed from weight fractions) are averaged into the new weight calculations to form the Class II weight estimate.

Step 8: The Class II empty weight of the Selene is 5,122 lbs. This compares with 4,900 lbs for the Class I weight estimate. This represents a difference of 222 lbs which is 4.5 percent of the Class I empty weight.

Several comments are in order:

1. an iteration through the equations of Step 7 should be performed, to determine the 'convergence' empty weight.

2. several weight savings can be made in the Selene:

a) by manufacturing the propellers out of composites, their weight can probably be cut by 40 percent for a weight saving of 93 lbs.

b) the empennage can be manufactured from composites which would yield a weight saving of about 15 percent, or 24 lbs.

c) the nacelles can be manufactured partially from composites which would yield a weight saving of about 10 percent, or 26 lbs.

d) by manufacturing the low stress areas of the wing and fuselage from composites, a weight saving of about 5 percent should be feasible. This would save 72 lbs.

e) combining a) through d) yields a saving of 215 lbs. It therefore appears quite possible to bring the overall Selene take-off weight in at the original estimate of 7,900 lbs.

Steps 9 and 10: Not needed, see item e), Step 8.

Step 11: To save space, this step has been omitted.

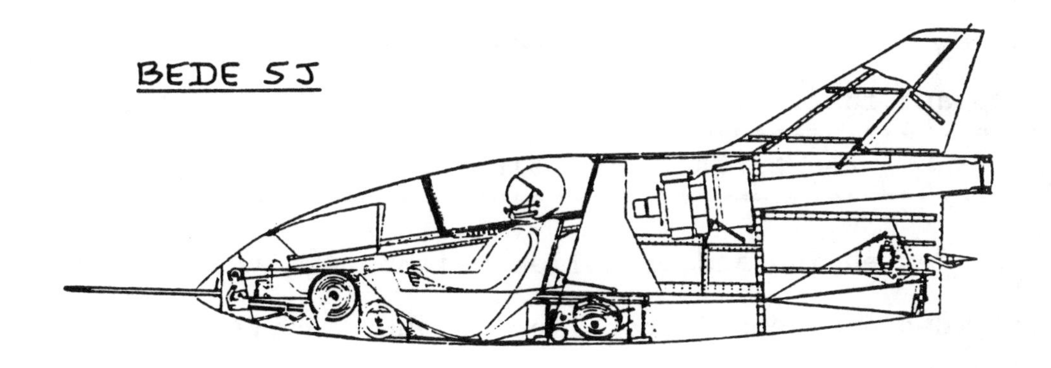

Table 4.1a Class II Weight Estimates for the Selene

Component	Methods: Class I Page 9	USAF	Torenbeek	Use as Class II Estimate
Structure weight, W_{struct}:				
Wing	738	580	410	576
Adjustment for Fowler flaps, 2 percent:				12
Empennage	179	149	155	161
Fuselage	621	830	1,130	860
Nacelles	249	N.A.	272	261
Landing gear	380	196	313	296
W_{struct}	2,167	1,755 excl.nac.	2,280	2,166
Powerplant weight, W_{pwr}:				
Engines	1,400	1,400	1,400	1,400
Air induction	in pwrplt		88	88
Propellers	200	250	250	233
Fuel system	in pwrplt	157	135	146
$W_{pwr}-W_{fs}$		2,162*	2,165**	
Powerplant inst.	108			108
W_{pwr}	1,708	2,319	2,300	1,975

* includes engine and propeller weight, Eqn.(6.3)

**includes engine and propeller weight, Eqn.(6.4)

Table 4.1b Class II Weight Estimates for the Selene

Component	Methods: Class I Page 9	Cessna	USAF	T'beek	Use as Class II Estimate
Fixed equipment weight, W_{feq}:					
W_{fc}		133	294	91	173
W_{hps}: this is included in W_{fc}					
W_{els}		212	210	209	210
W_{iae}				103	103
W_{api}				88	88
W_{ox}			GD: 25		25
W_{fur}		258		410	334
W_{pt}				Table A3.2a:	48
W_{feq}	1,025	not complete			981

Summary:

Class II empty weight, W_E follows from Eqn.(2.1):

$$W_E = 2,166 + 1,975 + 981 = 5,122 \text{ lbs}$$

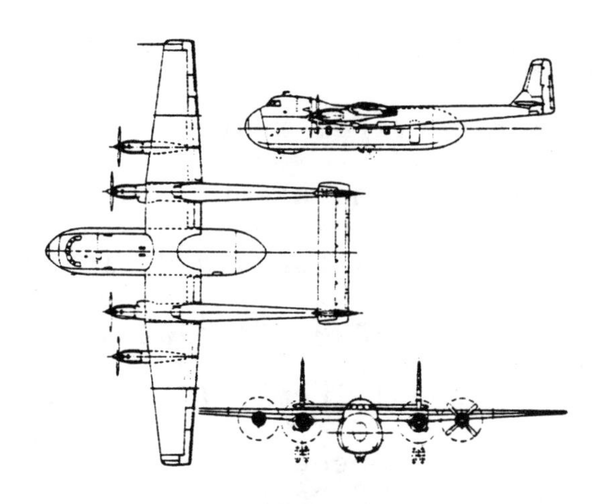

ARMSTRONG WHITWORTH ARGOSY 222

4.3.2 Jet Transport

Step 1: The following weight items are already known:

From Table 10.5, Part II:

Payload weight: W_{PL} = 30,750 lbs

Crew weight: W_{crew} = 1,025 lbs

Fuel weight: W_F = 25,850 lbs

Trapped fuel and oil: W_{tfo} = 925 lbs

From Part II, p.138:

Engine dry wht: W_e = 9,224 lbs

Step 2: Weights need to be estimated for the following items:

Structural Weight, W_{struct}:

1) Wing 2) Adjustment for Fowler flaps 3) Empennage

4) Fuselage 5) Nacelles 6) Landing Gear

Powerplant Weight, W_{pwr}:

1) Engines 2) Fuel system 3) Propulsion system

4) Accessory drives, starting and ignition system

5) Thrust reversers

Fixed Equipment Weight, W_{feq}:

1) Flight controls 2) Electrical system

3) Instrumentation, avionics and electronics

4) Air-conditioning, pressurization and de-icing

5) Oxygen 6) APU 7) Furnishings 8) Baggage and cargo handling 9) Operational items 10) Paint

Step 3: The structural arrangement drawing for the Ourania is presented in Chapter 8 of Part III.

Step 4: From a weight estimating viewpoint this airplane falls in the Commercial Transport category.

Step 5: The following weight equations apply to the Ourania:

W_{struct}: 1) Wing: Eqns (5.6) and (5.7)

2) Adjustment for Fowler flaps: an extra factor of 2 percent will be added in accordance with 5.2.2.2.

3) Empennage: Eqns (5.17), (5.18), (5.20)

4) Fuselage: Eqns (5.26) and (5.27)

5) Nacelles: Eqns (5.35) and (5.37)

6) Landing Gear: Eqns (5.41) and (5.42)

W_{pwr}: 1) Engines: see Step 1.

2) Fuel system: Eqn. (6.24)

3) Propulsion system: Eqns (6.24), (6.29)

4) Accessory drives, starting and ignition system: Eqn. (6.34)

5) Thrust reversers: Eqn. (6.36)

W_{feq}: 1) Flight control system: Eqns (7.5), (7.6)

Note: hydraulics and pneumatics are included in item 1).

2) Electrical system: Eqns (7.15), (7.17)

3) Instrumentation, avionics and electronics: Eqns (7.23) and (7.25)

4) Air-conditioning, pressurization and de-icing: Eqns (7.29) and (7.30)

5) Oxygen system: Eqns (7.35) and (7.37)

6) APU: Eqn.(7.40)

7) Furnishings: Eqns (7.44) and (7.45)

8) Baggage and cargo handling: Eqn.(7.48)

9) Operational items: See Section 7.10.

10) Paint: See Section 7.15.

Step 6: The following list itemizes all required input data for estimating the weight items listed in steps 2 and 5.

W_{TO} 127,000 lbs $n_{ult} = 2.5$ $S = 1,296\ ft^2$

$V_C = 295$ kts $V_D = 369$ kts $n_{ult} = 3.75$

$A = 10$ $\lambda = 0.32$ $\Lambda_{1/4} = 35$ deg.

$\Lambda_{1/2} = 33.5$ deg $M_H = 0.85$

$(t/c)_m = 0.13$ $b = 113.8$ ft $t_r = 2.26$ ft

$S_h = 254\ ft^2$ $b_h = 35.6$ ft $t_{r_h} = 1.30$ ft

$\bar{c} = 12.5$ ft $l_h = 32.5$ ft

$S_v = 200\ ft^2$ $z_h/b_v = 0$ $l_v = 35.8$ ft

$S_r/S_v = 0.45$ $\lambda_v = 0.32$ $A_v = 1.8$

$\Lambda_{1/4} = 45$ deg.

$l_f = 124.3$ ft $w_f + h_f = 26.4$ ft $\bar{q}_D = 461$ psf

$W_L = 7,505$ lbs $n_{ult.l.} = 4.0$ $d_f = 13.2$ ft

$P_2 = 20$ psi $l_n = 11.7$ ft $A_{inl} = 28.3\ ft^2$

$D_{inl} = 6.0$ ft

Notes: 1) The value for n_{lim} follows from the V-n diagram of Figure 4.5.

2) Most data were obtained from Ourania data listed in Part II. The reader is reminded that a detailed geometric definition may be found in Part II as Table 13.2, a Class I weight statement as Table 10.5. Detailed definitions of layouts of fuselage, powerplant, wing, high lift system, empennage and landing gear may be found in Chapters 4,5,6,7,8 and 9 respectively.

Step 7: Tables 4.2a - 4.2c list all weights computed as part of the Class II weight estimation process.

Step 8: The Class II empty weight of the Ourania is 72,622 lbs. This compares with 68,450 lbs for the Class I weight estimate. This represents a difference of 4,172 lbs which is 6.1 percent of the Class I empty weight.

Several comments are in order:

1. an iteration through the equations of Step 7 should be performed, to determine the 'convergence' empty weight.

2. several weight savings can be made in the Ourania:

a) the empennage can be manufactured from composites which would yield a weight saving of about 15 percent, or 359 lbs.

b) the nacelles can be manufactured partially from composites which would yield a weight saving of about 10 percent, or 264 lbs.

c) by manufacturing the low stress areas of the wing and fuselage from composites, a weight saving of about 5 percent should be feasible. This would save 1,388 lbs.

d) by using a quadruplex digital flight control system and using fly-by-wire instead of mechanical flight controls, a weight saving of 15 percent over the estimated weight can be obtained. This would save 352 lbs.

e) by using lithium aluminum in the primary wing and fuselage structure, a weight saving of 6 percent is feasible. This saves 1,665 lbs.

By combining a) through e) a total weight saving of 4,028 lbs can be achieved. This is close to the discrepancy of 4,172 lbs. It is therefore judged possible to bring the Ourania in at the originally estimated empty weight of 68,450 lbs.

Steps 9-10: Not needed, see item e), Step 8.

Step 11: This step has been omitted to save space.

Table 4.2a Class II Weight Estimates for the Ourania

Component	Methods: Class I Page 11	GD	Torenbeek	Use as Class II Estimate
Structure weight, W_{struct}:				
Wing	13,664	11,753	15,973	13,797
Adjustment for Fowler flaps, 2 percent:				276
Horiz.Tail		949	1,218	1,319
Vert.Tail		920	829	1,071
Empennage	3,253	1,869	2,047	2,390
Fuselage	14,184	15,748	11,140	13,691
Nacelles	2,082	2,722	3,120	2,641
Nose Gear	573		783	716
Main Gear	4,632		4,208	3,904
Landing gear	5,205	3,663	4,991	4,620
W_{struct}	38,388			37,415
Powerplant weight, W_{pwr}:				
Engines	9,224	9,224	9,224	9,224
Fuel system	in pwrplt		1,009	1,009
Propulsion inst.	667	439		700
Acc.dr, Start, Ign			960	
Thrust reversers			1,660	1,660
W_{pwr}	9,891			12,593

Table 4.2b Class II Weight Estimates for the Ourania:

Average Weight Fractions for Fixed Equipment Breakdown

Note: these data were used in Table 4.2c.

Component	Similar Airplane Type: McDD DC-9-30	MD-80	Boeing 737-200	727-100	Use as Class II Estimate
Fixed equipment weight item:					
W_{fc}*	0.0220	0.0241	0.0279	0.0276	0.0254
W_{els}	0.0123	0.0123	0.0092	0.0134	0.0118
W_{iae}	0.0134	0.0152	0.0137	0.0147	0.0143
W_{api}	0.0148	0.0152	0.0123	0.0124	0.0137
W_{ox}	0.0014	0.0016			0.0015
W_{apu}	0.0076	0.0060	0.0072		0.0069
W_{fur}	0.0782	0.0814	0.0575	0.0641	0.0703
W_{ops}	0.0250	0.0261			0.0256
W_{pt}	typical US airline paint scheme:				0.0035

* includes hydraulic and pneumatic system

Note: Specific airplane type data from Tables A7.1a and A7.2a in Appendix A.

Table 4.2c Class II Weight Estimates for the Ourania

Component	Methods: Table 4.2b x127,000	GD	Torenbeek	Use as Class II Estimate
Fixed equipment weight, W_{feq}:				
W_{fc}	3,226	2,200	1,617	2,348
W_{hps}: this is included in W_{fc}				
W_{els}	1,499	1,887	4,063	2,483
W_{iae}	1,810	1,593	1,775	1,726
W_{api}	1,737	4,251	2,166	2,718
W_{ox}	191	241	210	214
W_{apu}	881	1,016	1,016	982
W_{fur}	8,928	7,467	7,565	7,987
W_{bc}		466	466	466
W_{ops}	3,245		3,245	3,245
W_{pt}			Table 4.2b:	445
W_{feq}	21,517	19,121	22,123	22,614

Summary:

Class II empty weight, W_E follows from Eqn.(2.1):

$$W_E = 37,415 + 12,593 + 22,614 = 72,622 \text{ lbs}$$

4.3.3 Fighter

Step 1: The following weight items are already known:

From Table 10.6, Part II:

Payload weight: W_{PL} = 12,405 lbs

Crew weight: W_{crew} = 200 lbs

Fuel weight: W_F = 18,500 lbs

Trapped fuel and oil: W_{tfo} = 300 lbs

From Part II, p.140:

Engines, incl A/B: W_e = 6,000 lbs

Step 2: Weights need to be estimated for the following items:

Structural Weight, W_{struct}:

1) Wing 2) Adjustment for Fowler flaps 3) Empennage

4) Fuselage 5) Tailbooms 6) Engine section

7) Landing Gear

Powerplant Weight, W_{pwr}:

1) Engines 2) Afterburners 3) Air induction system

4) Fuel system 5) Propulsion system

Fixed Equipment Weight, W_{feq}:

1) Flight controls 2) Electrical system

3) Instrumentation, avionics and electronics

4) Air-conditioning, pressurization and de-icing

5) Armament 6) Furnishings 7) Oxygen system

8)Auxiliary gear 9) GAU-8A Gun

Step 3: The structural arrangement drawing for the Eris is presented in Chapter 8 of Part III.

Step 4: From a weight estimating viewpoint this airplane falls in the Fighter and Attack Airplane category.

Step 5: The following weight equations apply to the Eris:

W_{struct}: 1) Wing: Eqn. (5.9)

2) Adjustment for Fowler flaps: an extra factor of 2 percent will be added in accordance with 5.2.2.2.

3) Empennage: Eqns (5.17) and (5.18)

4) Fuselage: Eqn. (5.26)

5) Tailbooms: Eqn. (5.27)

6) Engine section: See Class I, p.14

7) Landing Gear: Eqns (5.41) and (5.42)

W_{pwr}: 1) Engines: see Step 1.

2) Air induction system: Eqn. (6.9)

3) Fuel system: Eqn. (6.20)

4) Propulsion system: Eqns (6.23), (6.27)

W_{feq}: The data of Table 4.3b are used, in addition to the following equations:

1) Flight control system: Eqn. (7.11)

Note: hydraulics and pneumatics are included in item 1).

2) Electrical system: Eqn. (7.19)

3) Instrumentation, avionics and electronics: Eqn. (7.25)

4) Air-conditioning, pressurization and de-icing: Eqn. (7.33)

5) Armament: Table 4.4b

6) Furnishings: Eqn. (7.47)

7) Oxygen system: Eqn. (7.39)

8) Auxiliary gear: Table 4.4b

9) GAU-8A Gun: See Part III under weapons

Step 6: The following list itemizes all required input data for estimating the weight items listed in steps 2 and 5.

$GW = 61{,}660$ lbs $\quad n_{ult} = 11.0 \quad S = 787\ ft^2$

$V_D = 563$ kts $\quad \bar{q}_D = 1{,}072$ psf $\quad n_{lim} = 7.33$

$A = 6 \quad K_w = 1.0 \quad \lambda = 0.50 \quad \Lambda_{LE} = 3.5$ deg.

$(t/c)_m = 0.10 \quad M_H = 0.68 \quad \bar{c} = 11.9$ ft

$S_h = 93\ ft^2 \quad b_h = 18.3$ ft $\quad t_{r_h} = 0.51$ ft

$l_h = 32.3$ ft

$S_v = 147^*\ ft^2 \quad z_h/b_v = 1.0 \quad l_v = 26.0$ ft

$S_r/S_v = 0.22 \quad \lambda_v = 0.55 \quad A_v = 1.2$

$\Lambda_{1/4} = 41$ deg. *This is for both vertical tails

$K_{inl} = 1.25$

$l_f = 41.3$ ft $\quad h_f = 6.83$ ft for the fuselage

$l_f = 33.3$ ft $\quad w_f + h_f = 3.06$ ft for the booms

$S_{fgs} = 2x30.6 = 61.2\ ft^2$ for the booms

	A_g	B_g	C_g	D_g
Nose gear:	12	0.06	0	0
Main gear:	33	0.04	0.021	0

$N_{inl} = 2 \quad L_d = 8$ ft $\quad A_{inl} = 6.31\ ft^2$

$P_2 = 30$ psi $\quad K_d = 1.0 \quad K_m = 1.0$

Notes: 1) The value for n_{lim} follows from the V-n diagram of Figure 4.6.

2) Most data were obtained from Eris data listed in Part II. The reader is reminded that a

detailed geometric definition may be found in Part II as Table 13.3, a Class I weight statement as Table 10.6. Detailed definitions of layouts of fuselage, powerplant, wing, high lift system, empennage and landing gear may be found in Chapters 4,5,6,7,8 and 9 respectively.

Step 7: Tables 4.3a, 4.3b and 4.3c list all weights computed as part of the Class II weight estimation process.

Step 8: The Class II empty weight of the Eris is 35,755 lbs. This compares with 33,500 lbs for the Class I weight estimate. This represents a difference of 2,255 lbs which is 6.7 percent of the Class I empty weight.

Several comments are in order:

1. an iteration through the equations of Step 7 should be performed, to determine the 'convergence' empty weight.

2. several weight savings can be made in the Eris:

a) the entire primary structure can be made from composites. This could yield a potential savings of 10 percent or 2,246 lbs.

b) by using a quadruplex digital flight control system and using fly-by-wire instead of mechanical flight controls, a weight saving of 15 percent over the estimated weight can be obtained. This would save 254 lbs.

By combining a) and b) a total weight saving of 2,500 lbs can be achieved. It is therefore judged possible to bring the Eris in at a weight below the originally estimated empty weight.

Steps 9-10: Not needed, see item e), Step 8.

Step 11: This step has been omitted to save space.

Table 4.3a Class II Weight Estimates for the Eris

Component	Methods: Class II Page 14	GD	Torenbeek	Use as Class II Estimate
Structure weight, W_{struct}:				
Wing	6,762	9,490		8,126
Adjustment for Fowler flaps, 2 percent:				163
Horiz.Tail		720		707
Vert.Tail		938		921
Empennage	1,597	1,658		1,628
Fuselage	7,347 incl.booms	5,044		5,967*
Booms		458		458
Engine Section	160			160
Nose Gear	554		267	443
Main Gear	2,214		1,603	1,768
Landing gear	2,768	1,996	1,870	2,211
W_{struct}	18,634	20,304		18,713
Powerplant weight, W_{pwr}:				
Engines	4,000	4,000	4,000	4,000
Afterburners	2,000	2,000	2,000	2,000
Air ind. syst.	in propuls.	445		445
Fuel system	in propuls.	777		777
Propulsion inst.	2,834**	78		845***
W_{pwr}	8,834	7,632		8,067

*1/2(7,347 - 458 + 5,044) = 5967

**includes air induction and fuel system

***1/2(2,834 + 78 - 445 - 777) = 845

Table 4.3b Weight Fraction Estimates for the Eris:

Average Weight Fractions for Fixed Equipment Breakdown

Note: these data were used in Table 4.3c.

Component	Similar Airplane Type: Republic F105B	Chance Vought F8U	Grumman A2F(A6)	Use as Class II Estimate
Fixed equipment weight item:				
W_{fc}*	0.0561	0.0515	0.0317	0.0464
W_{els}	0.0223	0.0144	0.0200	0.0189
W_{iae}	0.0307	0.0337	0.0800	0.0481
W_{api}	0.0054	0.0108	0.0047	0.0070
W_{arm}	0.0229	0.0123	0.0093	0.0148
W_{fur}	0.0077	0.0069	0.0137	0.0094
W_{aux}	0.0029	0.0060		0.0045

* includes hydraulic and pneumatic system

Note: Specific airplane type data from Tables A9.2a, A9.3a and A9.4a in Appendix A.

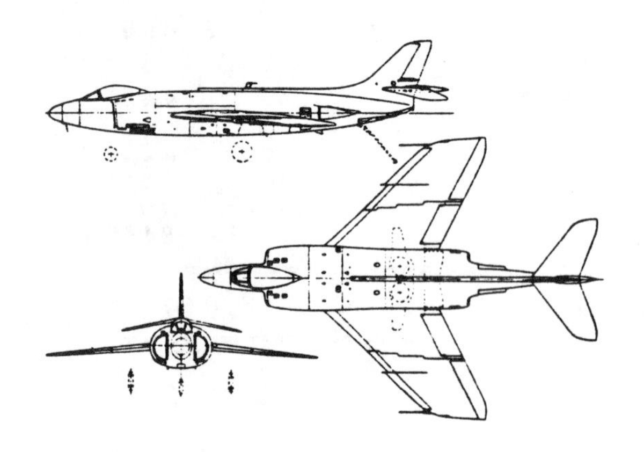

SUPERMARINE SCIMITAR F1

Table 4.3c Class II Weight Estimates for the Eris

Component	Methods: Table 4.3b x61,660	GD	Torenbeek	Use as Class II Estimate
Fixed equipment weight, W_{feq}:				
W_{fc}	3,459	1,513		2,486
$W_{fc_{cg}}$		102		102
W_{hps}: this is included in W_{fc}				
W_{els}	1,165	703		934
W_{iae}	1,893		1,033	1,463
W_{api}	431	347		389
W_{arm}	913			913
W_{fur}	580	214		397
W_{ox}	in W_{fur}	17		in W_{fur}
W_{aux}	277			277
GAU-8A Gun	2,014	Part II, Table 10.6:		2,014
W_{feq}	10,732*			8,975

* This disagrees significantly with W_{feq} in Table 10.6 of Part II.

Summary:

Class II empty weight, W_E follows from Eqn.(2.1):

$$W_E = 18,713 + 8,067 + 8,975 = 35,755 \text{ lbs}$$

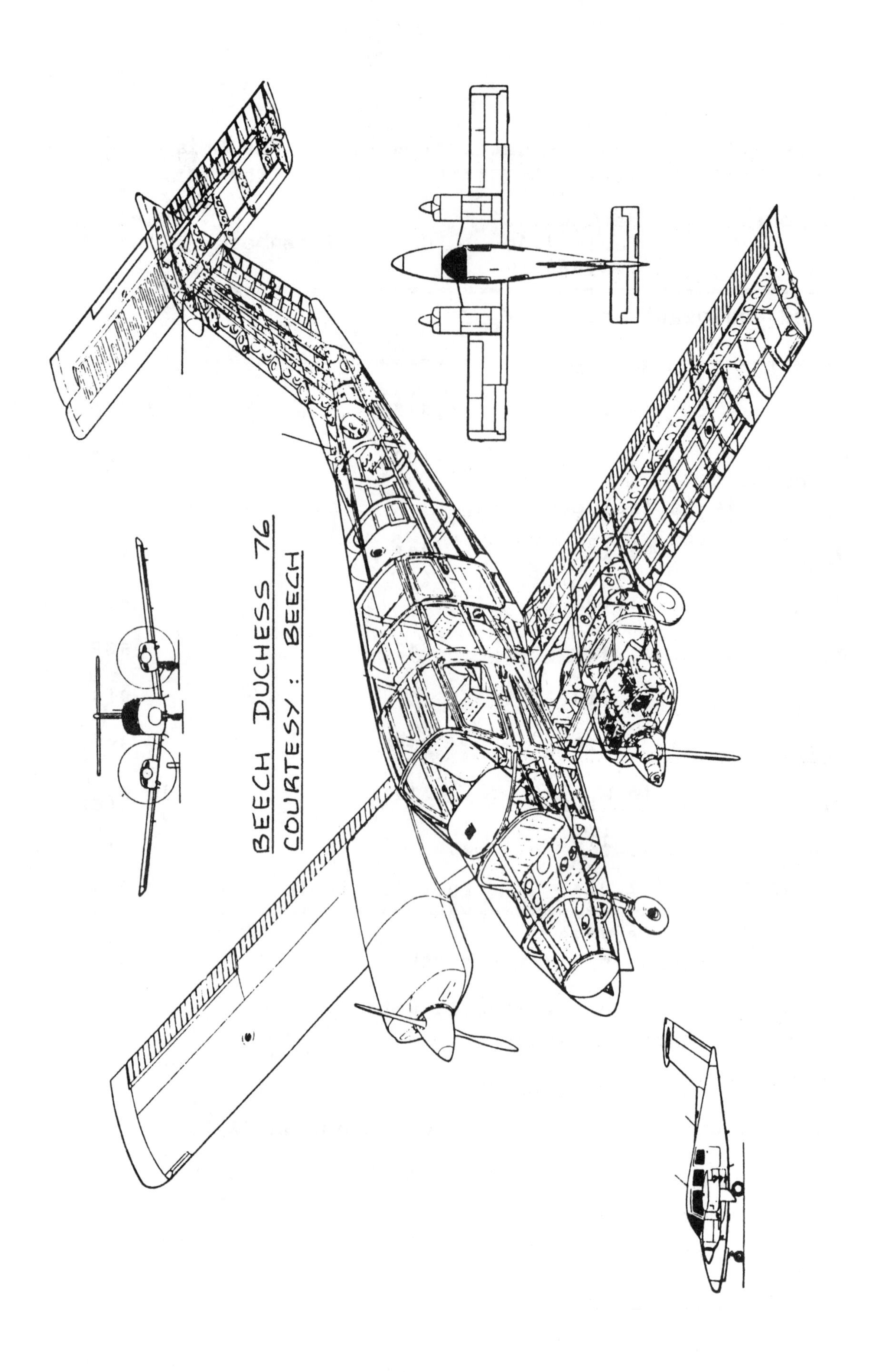
BEECH DUCHESS 76
COURTESY : BEECH

5. CLASS II METHOD FOR ESTIMATING STRUCTURE WEIGHT

The airplane structure weight, W_{struct} will be assumed to consist of the following components:

5.1 Wing, W_w 5.2 Empennage, W_{emp}

5.3 Fuselage, W_f 5.4 Nacelles, W_n

5.5 Landing gear, W_g Therefore:

$$W_{struct} = W_w + W_{emp} + W_f + W_n + W_g \quad (5.1)$$

Equations for structure weight estimation are presented for the following types of airplanes:

1. General Aviation Airplanes
2. Commercial Transport Airplanes
3. Military Patrol, Bomb and Transport Airplanes
4. Fighter and Attack Airplanes

5.1 WING WEIGHT ESTIMATION

5.1.1 General Aviation Airplanes

5.1.1.1 Cessna method

The following equations should be applied only to small, relatively low performance type airplanes with maximum speeds below 200 kts. The equations apply to wings of two types:

Cantilever wings: Eqn.(5.2)
Strut braced wings: Eqn.(5.3)

Both equations include: weight of wing tip fairing
wing control surfaces
Both equations exclude: fuel tanks
wing/fuselage spar carry-through structure
effect of sweep angle

For cantilever wings:

$$W_w = 0.04674(W_{TO})^{0.397}(S)^{0.360}(n_{ult})^{0.397}(A)^{1.712} \quad (5.2)$$

For strut braced wings:

$$W_w = 0.002933(S)^{1.018}(A)^{2.473}(n_{ult})^{0.611} \quad (5.3)$$

Definition of terms:

W_{TO} = take-off weight in lbs,

S = wing area in ft^2,

n_{ult} = design ultimate load factor

A = wing aspect ratio

Note that Eqn.(5.3) does not account for W_{TO}. It should therefore be used with caution. The reader should also realize that wings in this category have maximum thickness ratios of around 18 percent.

5.1.1.2 USAF Method

The following equation applies to light and utility type airplanes with performance up to about 300 kts:

$$W_w = 96.948[(W_{TO}n_{ult}/10^5)^{0.65}(A/\cos\Lambda_{1/4})^{0.57}(S/100)^{0.61} \times$$
$$\times\{(1+\lambda)/2(t/c)_m\}^{0.36}(1 + V_H/500)^{0.5}]^{0.993} \quad (5.4)$$

Definition of new terms:

$\Lambda_{1/4}$ = wing quarter chord sweep angle

λ = wing taper ratio

$(t/c)_m$ = maximum wing thickness ratio

V_H = maximum level speed at sealevel in kts

5.1.1.3 Torenbeek Method

The following equation applies to light transport airplanes with take-off weights below 12,500 lbs:

$$W_w =$$
$$= 0.00125W_{TO}(b/\cos\Lambda_{1/2})^{0.75}[1 + \{6.3\cos(\Lambda_{1/2})/b\}^{1/2}] \times$$
$$\times(n_{ult})^{0.55}(bS/t_rW_{TO}\cos\Lambda_{1/2})^{0.30} \quad (5.5)$$

See special notes in Section 5.2.2.

Definition of new terms:

b = wing span in ft

$\Lambda_{1/2}$ = wing semi-chord sweep angle

t_r = maximum thickness of wing root chord in ft

5.1.2 Commercial Transport Airplanes

5.1.2.1 GD Method

$$W_w = \frac{\{0.00428(S^{0.48})(A)(M_H)^{0.43}(W_{TO}n_{ult})^{0.84}(\lambda)^{0.14}\}}{[\{100(t/c)_m\}^{0.76}(\cos\Lambda_{1/2})^{1.54}]} \qquad (5.6)$$

Note: This equation is valid only in the following parameter ranges:

M_H from 0.4 to 0.8, $(t/c)_m$ from 0.08 to 0.15,

and A from 4 to 12.

Definition of new term:

M_H = maximum Mach number at sealevel

5.1.2.2 Torenbeek Method

The following equation applies to transport airplanes with take-off weights above 12,500 lbs:

$$W_w = 0.0017W_{MZF}(b/\cos\Lambda_{1/2})^{0.75}[1 + \{6.3\cos(\Lambda_{1/2})/b\}^{1/2}]\mathrm{x}$$
$$\mathrm{x}(n_{ult})^{0.55}(bS/t_rW_{MZF}\cos\Lambda_{1/2})^{0.30} \qquad (5.7)$$

Definition of new term:

W_{MZF} = maximum zero fuel weight = $W_{TO} - W_F$ (5.8)

Special notes:

1. Eqns.(5.6) and (5.7) include the weight of normal high lift devices as well as ailerons.
2. For spoilers and speed brakes 2 percent should be added.

3. If the airplane has 2 wing mounted engines reduce the wing weight by 5 percent.
4. If the airplane has 4 wing mounted engines reduce the wing weight by 10 percent.
5. If the landing gear is not mounted under the wing reduce the wing weight by 5 percent.
6. For braced wings reduce the wing weight by 30 percent. The resulting wing weight estimate does include the weight of the strut. The latter is roughly 10 percent of the wing weight.
7. For Fowler flaps add 2 percent to wing weight.

5.1.3 Military Patrol, Bomb and Transport Airplanes

For predicting wing weight it is suggested to use Eqns.(5.6) and (5.7) but with the appropriate value for n_{ult}. For this type of military airplane the usual value for n_{ult} is 4.5. Refer to Table 4.1 for a listing of military limit load factors.

Note: wing weight in military airplanes is often based on the flight design gross weight, GW, rather than W_{TO}. Check the mission specification and/or the applicable military specifications to determine which weight value to use in Eqns.(5.6) and (5.7).

5.1.4 Fighter and Attack Airplanes

5.1.4.1 GD Method

For USAF fighter and attack airplanes:

$$W_w =$$

$$= 3.08[\{(K_w n_{ult} W_{TO})/(t/c)_m\}\{(\tan\Lambda_{LE} - 2(1-\lambda)/A(1+\lambda))^2 + $$
$$+ 1.0\}\times 10^{-6}]^{0.593}\{A(1+\lambda)\}^{0.89}(S)^{0.741} \quad (5.9)$$

For USN fighter and attack airplanes:

$$W_w =$$

$$= 19.29[\{(K_w n_{ult} W_{TO})/(t/c)_m\}\{(\tan\Lambda_{LE} - 2(1-\lambda)/A(1+\lambda))^2 + $$
$$+ 1.0\}\times 10^{-6}]^{0.464}\{(1+\lambda)A\}^{0.70}(S)^{0.58} \quad (5.10)$$

Definition of new terms:

K_w = 1.00 for fixed wing airplanes and
= 1.175 for variable sweep wing airplanes

Λ_{LE} = leading edge sweep angle of the wing

Note: wing weight in military airplanes is often based on the flight design gross weight, GW, rather than W_{TO}. Check the mission specification and/or the applicable military specifications to determine which weight to use in Eqns.(5.9) and (5.10).

5.2 EMPENNAGE WEIGHT ESTIMATION

Empennage weight, W_{emp} will be expressed as follows:

$$W_{emp} = W_h + W_v + W_c, \tag{5.11}$$

where: W_h = horizontal tail weight in lbs

W_v = vertical tail weight in lbs

W_c = canard weight in lbs

Equations for empennage weight components are presented in the remainder of this section.

5.2.1 General Aviation Airplanes

5.2.1.1 Cessna method

The following equations should be applied only to small, relatively low performance type airplanes with maximum speeds below 200 kts.

Horizontal tail:

$$W_h = \frac{3.184(W_{TO})^{0.887}(S_h)^{0.101}(A_h)^{0.138}}{174.04(t_{r_h})^{0.223}} \tag{5.12}$$

Note that no factor for horizontal tail sweep is included.

Vertical tail:

$$W_v = \frac{1.68(W_{TO})^{0.567}(S_v)^{1.249}(A_v)^{0.482}}{639.95(t_{r_v})^{0.747}(\cos\Lambda_{1/4_v})^{0.882}} \tag{5.13}$$

Canard: For a lightly loaded canard, Eqn.(5.12) may be used. For a significantly loaded canard (such as on the GP180 and the Starship I) it is suggested to use the appropriate wing weight equation.

Definition of terms:

W_{TO} = take-off weight in lbs

S_h = horizontal tail area in ft^2

A_h = horizontal tail aspect ratio

t_{r_h} = horizontal tail maximum root thickness in ft

S_v = vertical tail area in ft^2

A_v = vertical tail aspect ratio

t_{r_v} = vertical tail maximum root thickness in ft

$\Lambda_{1/4_v}$ = vertical tail quarter chord sweep angle

5.2.1.2 USAF Method

The following equation applies to light and utility type airplanes with performance up to about 300 kts:

Horizontal tail:

$$W_h = 127\{(W_{TO}n_{ult}/10^5)^{0.87}(S_h/100)^{1.2} \times 0.289(l_h/10)^{0.483}(b_h/t_{r_h})^{0.5}\}^{0.458} \quad (5.14)$$

Note that sweep angle is not a factor in this equation.

Vertical tail:

$$W_v = 98.5\{(W_{TO}n_{ult}/10^5)^{0.87}(S_v/100)^{1.2} \times 0.289(b_v/t_{r_v})^{0.5}\}^{0.458} \quad (5.15)$$

Again, sweep angle is not a factor in this equation.

Canard:

The comments made under 5.2.1.1 also apply.

Definition of new terms:

l_h = distance from wing $\bar{c}/4$ to hor. tail $\bar{c}_h/4$ in ft

b_h = horizontal tail span in ft

b_v = vertical tail span in ft

5.2.1.3 Torenbeek Method

The following equation applies to light transport airplanes with design dive speeds up to 250 kts and with conventional tail configurations:

$$W_{emp} = 0.04\{n_{ult}(S_v + S_h)^2\}^{0.75}, \tag{5.16}$$

If the airplane also has a canard, the comments made under 'canard' in 5.2.1.1 also apply here.

5.2.2 Commercial Transport Airplanes

5.2.2.1 GD Method

Horizontal tail:

$$W_h = 0.0034\{(W_{TO}n_{ult})^{0.813}(S_h)^{0.584} \times (b_h/t_{r_h})^{0.033}(\bar{c}/l_h)^{0.28}\}^{0.915} \tag{5.17}$$

Note: sweep angle is not a factor in this equation.

Vertical tail:

$$W_v = 0.19\{(1 + z_h/b_v)^{0.5}(W_{TO}n_{ult})^{0.363}(S_v)^{1.089}(M_H)^{0.601} \times (l_v)^{-0.726}(1 + S_r/S_v)^{0.217}(A_v)^{0.337}(1+\lambda_v)^{0.363} \times (\cos\Lambda_{1/4_v})^{-0.484}\}^{1.014} \tag{5.18}$$

Canard: Comments made under 5.2.1.1 also apply here.

Definition of new terms:

z_h = distance from the vertical tail root to where the horizontal tail is mounted on the vertical tail, in ft. Warning: for fuselage mounted horizontal tails, set z_h = 0.

l_v = dist. from wing $\bar{c}/4$ to vert. tail $\bar{c}_v/4$ in ft

S_r = rudder area in ft^2

λ_v = vertical tail taper ratio

5.2.2.2 Torenbeek Method

The following equation applies to transport airplanes and to business jets with design dive speeds above 250 kts.

Horizontal tail:

$$W_h = \tag{5.19}$$

$$= K_h S_h [3.81\{(S_h)^{0.2} V_D\}/\{1,000(\cos \Lambda_{1/2_h})^{1/2}\} - 0.287]$$

where K_h takes on the following values:

K_h = 1.0 for fixed incidence stabilizers

K_h = 1.1 for variable incidence stabilizers

Vertical tail:

$$W_v = \tag{5.20}$$

$$= K_v S_v [3.81\{(S_v)^{0.2} V_D/\{1,000(\cos \Lambda_{1/2_v})^{1/2}\} - 0.287]$$

where K_v takes on the following values:

K_v = 1.0 for fuselage mounted horizontal tails

for fin mounted horizontal tails:

$$K_v = \{1 + 0.15(S_h z_h / S_v b_v)\} \tag{5.21}$$

Definition of new terms:

V_D = design dive speed in KEAS

$\Lambda_{1/2_h}$ = horizontal tail semi-chord sweep angle

$\Lambda_{1/2_v}$ = vertical tail semi-chord sweep angle

Canard: The comments made under 5.2.1.1 also apply here.

5.2.3 Military Patrol, Bomb and Transport Airplanes

See Sub-section 5.2.4.

5.2.4 Fighter and Attack airplanes

For estimation of empennage weight of airplanes in this category, use the methods of sub-section 5.2.2. Be sure to use the proper values for ultimate load factor. See Table 4.1.

Note: empennage weights of military airplanes are often based on the flight design gross weight, GW, rather than W_{TO}. Check the mission specification and/or the applicable military specifications to determine which weight to use.

5.3 FUSELAGE WEIGHT ESTIMATION

The equations presented for fuselage weight estimation are valid for land-based airplanes only. For flying boats and amphibious airplanes it is suggested to multiply the fuselage weight by 1.65:

$$W_{f_{fl.boat}} = 1.65W_f \quad (5.22)$$

For float equipped airplanes the weight due to the floats may be found with Eqn.(5.27), by substituting float wetted area for S_{fgs}.

For estimation of tailboom weight it is suggested to use Eqn.(5.27) applied to each tailboom individually, but with $K_f = 1$.

5.3.1 General Aviation airplanes

5.3.1.1 Cessna method

The following equations should be applied only to small, relatively low performance type airplanes with maximum speeds below 200 kts.

For low wing airplanes:

$$W_f = \quad (5.23)$$

$$= 0.04682(W_{TO})^{0.692}(P_{max})^{0.374}(l_{f-n})^{0.590}$$

For high wing airplanes:

$$W_f = \tag{5.24}$$

$$14.86(W_{TO})^{0.144}(l_{f-n}/p_{max})^{0.778}(l_{f-n})^{0.383}(N_{pax})^{0.455}$$

Definition of terms:

W_{TO} = take-off weight in lbs

N_{pax} = number of passengers including the pilots

l_{f-n} = fuselage length, not including nose mounted nacelle length in ft

P_{max} = maximum fuselage perimeter in ft

Notes: 1. These equations do not account for pressurized fuselages.
2. For this type airplane the crew is counted in the number of passengers.

5.3.1.2 USAF Method

The following equation applies to light and utility type airplanes with performance up to about 300 kts:

$$W_f = 200[(W_{TO}n_{ult}/10^5)^{0.286}(l_f/10)^{0.857} x$$

$$x\{(w_f + h_f)/10\}(V_C/100)^{0.338}]^{1.1} \tag{5.25}$$

Definition of new terms:

n_{ult} = ultimate load factor

l_f = fuselage length in ft

w_f = maximum fuselage width in ft

h_f = maximum fuselage height in ft

V_C = design cruise speed in KEAS

5.3.2 Commercial Transport Airplanes

5.3.2.1 GD Method

$$W_f = \tag{5.26}$$

$$= 10.43(K_{inl})^{1.42}(\bar{q}_D/100)^{0.283}(W_{TO}/1000)^{0.95}(l_f/h_f)^{0.71}$$

The factor K_{inl} takes on the following values:

K_{inl} = 1.25 for airplanes with inlets in or on the fuselage for a buried engine installation

K_{inl} = 1.0 for inlets located elsewhere

Definition of new term:

$\bar{q}_D$ = design dive dynamic pressure in psf

5.3.2.2 Torenbeek Method

The following equation applies to transport airplanes and to business jets with design dive speeds above 250 kts.

$$W_f = 0.021K_f\{(V_D l_h/(w_f + h_f)\}^{1/2}(S_{fgs})^{1.2} \qquad (5.27)$$

The constant K_f takes on the following values:

K_f = 1.08 for a pressurized fuselage

= 1.07 for a main gear attached to the fuselage.

= 1.10 for a cargo airplane with a cargo floor

These effects are multiplicative for airplanes equipped with all of the above.

Definition of new terms:

V_D = design dive speed in KEAS

l_h = distance from wing root c/4 to hor. tail root c/4 in ft

S_{fgs} = fuselage gross shell area in ft^2

5.3.3 Military Patrol, Bomb and Transport Airplanes

5.3.3.1 GD Method

For USAF airplanes, Eqn.(5.26) may be used.

For USN airplanes the following equation should be used:

$$W_f = \qquad (5.28)$$

$$= 11.03(K_{inl})^{1.23})(\bar{q}_L/100)^{0.245}(W_{TO}/1000)^{0.98}(l_f/h_f)^{0.61}$$

Values for K_{inl} are as given in 5.3.2.1.

Definition of new term:

$\bar{q}_L$ = design dive dynamic pressure in psf

5.3.4 Fighter and Attack Airplanes

For estimation of fuselage weights Equations (5.26) or (5.28) may be used.

5.4 NACELLE WEIGHT ESTIMATION

The nacelle weight is assumed to consist of the following components:

1. For podded engines: the structural weight associated with the engine external ducts and or cowls. Any pylon weight is included.

2. For propeller driven airplanes: the structural weight associated with the engine external ducts and or cowls plus the weight due to the engine mounting trusses.

3. For buried engines: the structural weight associated with special cowling and or ducting provisions (other than the inlet duct which is included in the air induction system under powerplant weight, Section 6.2) and any special engine mounting provisions.

5.4.1 General Aviation Airplanes

5.4.1.1 Cessna Method

The following equations should be applied only to small, relatively low performance type airplanes with maximum speeds below 200 kts.

$$W_n = K_n P_{TO} \tag{5.29}$$

The constant K_n takes on the following values:

K_n = 0.37 lbs/hp for radial engines

K_n = 0.24 lbs/hp for horizontally opposed engines

Definition of term:

P_{TO} = take-off power in hp

These data should not be applied to turbopropeller nacelles.

5.4.1.2 USAF Method

In this method, the nacelle weight is included in the powerplant weight: refer to Chapter 6.

5.4.1.3 Torenbeek Method

For single engine propeller driven airplanes with the nacelle in the fuselage nose:

$$W_n = 2.5(P_{TO})^{1/2} \quad (5.30)$$

This weight includes the entire engine section forward of the firewall.

For multi-engine airplane with piston engines:

$$W_n = 0.32 P_{TO} \text{ for horizontally opposed engines} \quad (5.31)$$

$$W_n = 0.045(P_{TO})^{5/4}(N_e)^{-1/4} \text{ for radial engines} \quad (5.32)$$

$$W_n = 0.14(P_{TO}) \text{ for turboprop engines} \quad (5.33)$$

Notes:

1. Since P_{TO} is the total required take-off horsepower, these weight estimates include the weights of all nacelles.
2. If the main landing gear retracts into the nacelles, add 0.04 lbs/hp to the nacelle weight
3. If the engine exhausts over the wing, as in the Lockheed Electra, add 0.11 lbs/hp to the nacelle weight.

5.4.2 Commercial Transport Airplanes

5.4.2.1 GD Method

For turbojet engines:

$$W_n = 3.0(N_{inl})\{(A_{inl})^{0.5}(l_n)(P_2)\}^{0.731} \quad (5.34)$$

For turbofan engines:

$$W_n = 7.435(N_{inl})\{(A_{inl})^{0.5}(l_n)(P_2)\}^{0.731} \tag{5.35}$$

Definition of terms:

N_{inl} = number of inlets

A_{inl} = capture area per inlet inft^2

l_n = nacelle length from inlet lip to compressor face in ft

P_2 = maximum static pressure at engine compressor face in psi. Typical values range from 15 to 50 psi.

5.4.2.2 Torenbeek Method

For turbojet or low bypass ratio turbofan engines:

$$W_n = 0.055T_{TO} \tag{5.36}$$

For high bypass ratio turbofan engines:

$$W_n = 0.065T_{TO} \tag{5.37}$$

Since T_{TO} is the total required take-off thrust, these equations account for the weight of all nacelles.

5.4.3 Military Patrol Bomb and Transport Airplanes

For all airplanes in this category Eqns.(5.34) and (5.35) may be used.

5.4.4 Fighter and Attack Airplanes

For all airplanes in this category Eqns.(5.34) and (5.35) may be used.

5.5 LANDING GEAR WEIGHT ESTIMATION

5.5.1 General Aviation Airplanes

5.5.1.1 Cessna method

The following equations should be applied only to small, relatively low performance type airplanes with maximum speeds below 200 kts.

For non-retractable landing gears:

$$W_g = \qquad (5.38)$$

$$0.013W_{TO} + 0.362(W_L)^{0.417}(n_{ult.l})^{0.950}(l_{s_m})^{0.183} +$$

wheels + tires m.g. strut assembly m.g.

$$+ 6.2 + 0.0013W_{TO} + 0.007157(W_L)^{0.749}(n_{ult.l})(l_{s_n})^{0.788}$$

wheels + tires n.g. strut assembly n.g.

For retractable landing gears:

$$W_g = W_{g_{Eqn.(5.38)}} + 0.014W_{TO} \qquad (5.39)$$

Definition of terms:

W_{TO} = take-off weight in lbs

W_L = design landing weight in lbs (See Table 3.3, Part I for data relating W_L to W_{TO})

$n_{ult.l}$ = ultimate load factor for landing, may be taken as 5.7

l_{s_m} = shock strut length for main gear in ft

l_{s_n} = shock strut length for nose gear in ft

5.5.1.2 USAF Method

The following equation applies to light and utility type airplanes with performance up to about 300 kts:

$$W_g = 0.054(l_{s_m})^{0.501}(W_L n_{ult.l})^{0.684} \qquad (5.40)$$

Notes: 1) This equation includes nose gear weight.

2) $n_{ult.l}$ may be taken as 5.7.

5.5.2 Commercial Transport Airplanes

5.5.2.1 GD Method

$$W_g = 62.21(W_{TO}/1,000)^{0.84} \qquad (5.41)$$

5.5.2.2 Torenbeek Method

The following equation applies to transport airplanes and to business jets with the main gear mounted on the wing and the nose gear mounted on the fuselage:

$$W_g = K_{g_r}\{A_g + B_g(W_{TO})^{3/4} + C_g W_{TO} + D_g(W_{TO})^{3/2}\} \qquad (5.42)$$

The factor K_{g_r} takes on the following values:

K_{g_r} = 1.0 for low wing airplanes

K_{g_r} = 1.08 for high wing airplanes

The constants A_g through D_g are defined in Table 5.1 which is taken from Reference 14.

Table 5.1 Constants in Landing Gear Weight Eqn.(5.42)

Airplane Type	Gear Type	Gear Comp.	A_g	B_g	C_g	D_g
Jet Trainers and Business Jets	Retr.	Main	33.0	0.04	0.021	0.0
		Nose	12.0	0.06	0.0	0.0
Other civil airplanes	Fixed	Main	20.0	0.10	0.019	0.0
		Nose	25.0	0.0	0.0024	0.0
		Tail	9	0.0	0.0024	0.0
	Retr.	Main	40.0	0.16	0.019	1.5×10^{-5}
		Nose	20.0	0.10	0.0	2.0×10^{-6}
		Tail	5.0	0.0	0.0031	0.0

5.5.3 Military Patrol, Bomb and Transport Airplanes

For USAF airplanes, Eqns.(5.41) and (5.42) may be used.

For USN airplanes the following equation should be used:

$$W_g = 129.1(W_{TO}/1{,}000)^{0.66} \qquad (5.43)$$

5.5.4 Fighter and Attack Airplanes

For USAF airplanes, Eqns.(5.41) and (5.42) may be used.

For USN airplanes, Eqn.(5.43) should be used.

6. CLASS II METHOD FOR ESTIMATING POWERPLANT WEIGHT

The airplane powerplant weight, W_{pwr} will be assumed to consist of the following components:

6.1 Engines, W_e: this includes engine, exhaust, cooling, supercharger and lubrication systems.

<u>Note:</u> afterburners and thrust reversers are not always included under engines. They are often treated as a separate powerplant component.

6.2 Air induction system, W_{ai}: this includes inlet ducts other than nacelles, ramps, spikes and associated controls.

6.3 Propellers, W_{prop}

6.4 Fuel System, W_{fs}

6.5 Propulsion System, W_p, this includes:

*engine controls
*starting systems
*propeller controls
*provisions for engine installation

<u>Note:</u> instead of the words 'propulsion system', the words 'propulsion installation' or even 'engine installation' are sometimes used.

Therefore:

$$W_{pwr} = W_e + W_{ai} + W_{prop} + W_{fs} + W_p \qquad (6.1)$$

<u>General Note:</u> for powerplant weight predictions it is highly recommended to obtain actual weight data from engine manufacturers.

Equations for powerplant weight prediction are presented for the following types of airplanes:

1. General Aviation Airplanes
2. Commercial Transport Airplanes
3. Military Patrol, Bomb and Transport Airplanes
4. Fighter and Attack Airplanes

6.1 ENGINE WEIGHT ESTIMATION

6.1.1 General Aviation Airplanes

6.1.1.1 Cessna method

The following equations should be applied only to small, relatively low performance type airplanes with maximum speeds below 200 kts.

$$W_e = K_p P_{TO} \qquad (6.2)$$

The factor K_p takes on the following values:

For piston engines: K_p = 1.1 to 1.8, depending on whether or not supercharging is used.

For turbopropeller engines: K_p = 0.35 to 0.55.

These weights represent the so-called engine dry weight. Normal engine accessories are included in this weight but engine oil is not.

Definition of terms:

W_e = weight of all engines in lbs

P_{TO} = required take-off power in hp

6.1.1.2 USAF Method

$$W_e + W_{ai} + W_{prop} + W_p = 2.575(W_{eng})^{0.922}N_e \qquad (6.3)$$

Use engine manufacturers data to obtain W_{eng} or use Eqn.(6.2).

Definition of new terms:

W_{eng} = weight per engine in lbs

N_e = number of engines

6.1.1.3 Torenbeek Method

For propeller driven airplanes:

$$W_{pwr} = K_{pg}(W_e + 0.24P_{TO}) \qquad (6.4)$$

The constant K_{pg} takes on the following values:

K_{pg} = 1.16 for single engine tractor installations

K_{pg} 1.35 for multi-engine installations

For superchargers the following additional weight is incurred:

$$W_{sprch} = 0.455(W_e)^{0.943} \tag{6.5}$$

For jet airplanes:

$$W_{pwr} = K_{pg}K_{thr}W_e \tag{6.6}$$

The constant K_{pg} takes on the following values:

K_{pg} = 1.40 for airplanes with buried engines

The constant K_{thr} takes on the following values:

K_{thr} = 1.00 for airplanes without thrust reversers

K_{thr} = 1.18 for airplanes with thrust reversers

6.1.2 Commercial Transport Airplanes

Use of actual engine manufactures data is highly recommended. Figure 6.1 provides a graphical summary of engine dry weights versus take-off thrust. Figure 6.2 gives a graphical summary of engine dry weights versus take-off shaft horsepower.

When using Figures (6.1) or (6.2), keep in mind that:

$$W_e = N_e W_{eng}, \tag{6.7}$$

where W_{eng} is the weight per engine.

Equations (6.4) and (6.6) may also be used to obtain an initial estimate.

6.1.3 Military Patrol, Bomb and Transport Airplanes

See Sub-Section 6.1.2.

6.1.4 Fighter and Attack Airplanes

See Sub-Section 6.1.2.

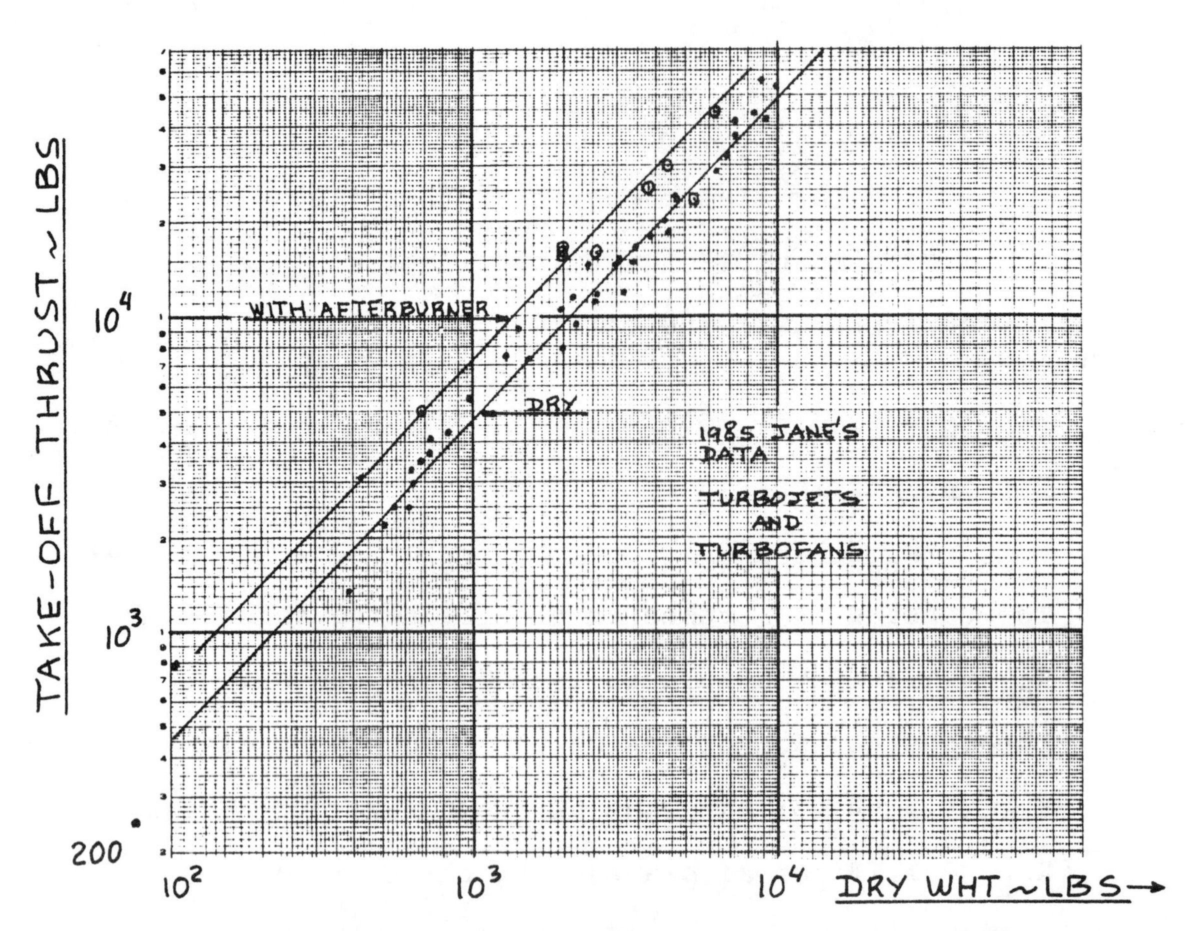

Figure 6.1 Turbojets and Turbofans: Take-off Thrust and Dry Weight Trends

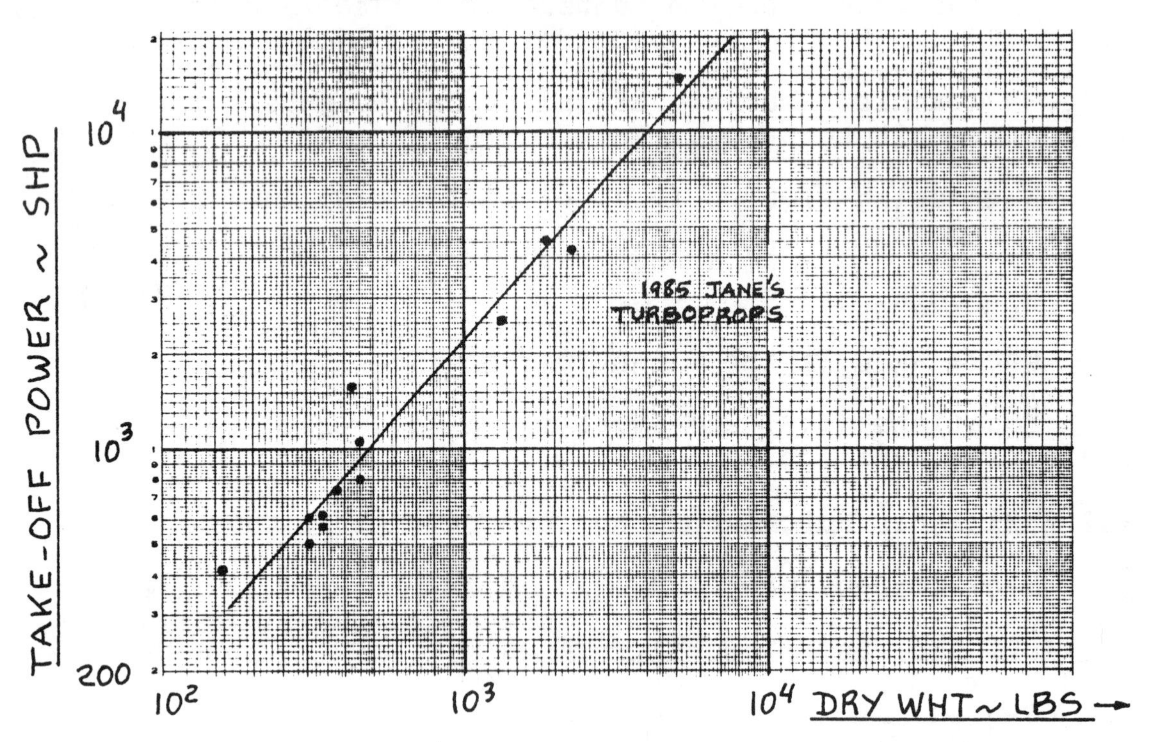

Figure 6.2 Turboprops: Take-off Shaft Horse Power and Dry Weight Trends

6.2 AIR INDUCTION SYSTEM WEIGHT ESTIMATION

6.2.1 General Aviation Airplanes

6.2.1.1 Cessna method

W_{ai} is included in the propulsion system weight, W_p.

6.2.1.2 USAF Method

See 6.2.1.1.

6.2.1.3 Torenbeek Method

$$W_{ai} + W_p = 1.03(N_e)^{0.3}(P_{TO})^{0.7} \qquad (6.8)$$

6.2.2 Commercial Transport Airplanes

6.2.2.1 GD Method

For buried engine installations:

The air induction system weight is split into two items: the first one for duct support structure, the second one for the subsonic duct leading from the inlet lip to the engine compressor face.

$$W_{ai} = 0.32(N_{inl})(L_d)(A_{inl})^{0.65}(P_2)^{0.6} + \qquad (6.9)$$

(duct support structure)

$$1.735\{(L_d)(N_{inl})(A_{inl})^{0.5}(P_2)(K_d)(K_m)\}^{0.7331}$$

(subsonic part of duct)

The factors K_d and K_m are defined as follows:

K_d = 1.33 for ducts with flat cross sections

1.0 for ducts with curved cross sections

K_m = 1.0 for M_D below 1.4

= 1.5 for M_D above 1.4

Definition of terms:

L_d = duct length in ft

N_{inl} = number of inlets

A_{inl} = capture area per inlet in ft^2

P_2 = maximum static pressure at engine compressor face in psi. Typical values range from 15 to 50 psi.

For podded engine installations:

The air induction system weight is included in the nacelle weight, W_n.

6.2.2.2 Torenbeek Method

For buried engine installations:

$$W_{ai} = 11.45\{(L_d)(N_{inl})(A_{inl})^{0.5}(K_d)\}^{0.7331} \qquad (6.10)$$

The constant K_d takes on the following values:

K_d = 1.0 for ducts with curved cross sections

1.33 for ducts with flat cross sections

For podded engine installations:

The air induction system weight is included in the nacelle weight, W_n.

Note: For supersonic installations additional weight items due to the special inlet requirements are needed. See Sub-section 6.2.4.

6.2.3. Military Patrol, Bomb and Transport Airplanes

See Section 6.2.2.

6.2.4 Fighter and Attack Airplanes

6.2.4.1 GD Method

For prediction of the duct support structure weight and the duct weight, Eqn.(6.9) may be used.

Particularly in supersonic applications the following additional weight items due to inlet provisions may be incurred:

For variable geometry ramps, actuators and controls:

$$W_{ramp} = 4.079\{(L_r)(N_{inl})(A_{inl})^{0.5}(K_r)\}^{1.201} \qquad (6.11)$$

The factor K_r takes on the following values:

K_r = 1.0 for M_D below 3.0

= $(M_D + 2)/5$ for M_D above 3.0

Definition of new term:

L_r is the ramp length forward of the inlet throat in ft

For inlet spikes:

$$W_{sp} = K_s(N_{inl})(A_{inl}) \quad (6.12)$$

The constant K_s takes on the following values:

K_s = 12.53 for half round fixed spikes
= 15.65 for full round translating spikes
= 51.80 for translating and expanding spikes

Note: these weights also apply to supersonic commercial installations.

6.3 PROPELLER WEIGHT ESTIMATION

6.3.1 General Aviation Airplanes

It is recommended to use propeller manufacturer data whereever possible. Lacking actual data the equation of Sub-Section 6.3.2 may be used.

Appendix A contains propeller installation data for a number of airplanes. Propeller installation weights usually include the propeller controls.

6.3.2 Commercial Transport Airplanes

6.3.2.1 GD Method

$$W_{prop} = K_{prop1}(N_p)(N_{bl})^{0.391}\{(D_p)(P_{TO}/N_e)/1{,}000\}^{0.782} \quad (6.13)$$

The constant K_{prop1} takes on the following values:

K_{prop1} = 24.0 for turboprops above 1,500 shp

= 31.92 for piston engines and for turbo-props below 1,500 shp

Definition of terms:

N_p is the number of propellers

N_{bl} is the number of blades per propeller

D_p is the propeller diameter in ft

P_{TO} is the required take-off power in hp

N_e is the number of engines

6.3.2.2 Torenbeek Method

$$W_{prop} = K_{prop2}(N_p)^{0.218}\{D_p P_{TO}(N_{bl})^{1/2}\}^{0.782} \quad (6.14)$$

The factor K_{prop2} takes on the following values:

K_{prop2} = 0.108 for turboprops

K_{prop2} = 0.144 for piston engines

The reader is asked to show that equations (6.13) and (6.14) are in fact the same.

6.3.3 Military Patrol,Bomb and Transport Airplanes

See Sub-Section 6.3.2.

6.3.4 Fighter and Attack Airplanes

See Sub-Section 6.3.2.

6.4 FUEL SYSTEM WEIGHT ESTIMATION

Note: In some airplanes the fuel system is used to control the center of gravity location. Airplanes with relaxed static stability and/or supersonic cruise airplanes frequently require such a system. The weight increment incurred due to such a feature is included in the weight estimation of the flight control system, Section 7.1.

6.4.1 General Aviation Airplanes

6.4.1.1 Cessna method

For airplanes with internal fuel systems (no tiptanks):

$$W_{fs} = 0.40W_F/K_{fsp} \quad (6.15)$$

For airplanes with external fuel systems (with tiptanks):

$$W_{fs} = 0.70 W_F / K_{fsp} \quad (6.16)$$

The constant K_{fsp} takes on the following values:

K_{fsp} = 5.87 lbs/gal for aviation gasoline

= 6.55 for lbs/gal for JP-4

Definition of term:

W_F = mission fuel weight (includes reserves) in lbs

6.4.1.2 USAF Method

$$W_{fs} = \quad (6.17)$$

$$= 2.49[(W_F/K_{fsp})^{0.6}\{1/(1+int)\}^{0.3}(N_t)^{0.20}(N_e)^{0.13}]^{1.21}$$

The factor K_f is defined in 6.4.1.1.

Definition of new terms:

int = fraction of fuel tanks which are integral

N_t = number of separate fuel tanks

N_e = number of engines

6.4.1.3 Torenbeek Method

For turbine engines, see Sub-Section 6.4.2.

For single piston engine installations:

$$W_{fs} = 2(W_F/5.87)^{0.667} \quad (6.18)$$

For multi piston engine installations:

$$W_{fs} = 4.5(W_F/5.87)^{0.60} \quad (6.19)$$

6.4.2 Commercial Transport Airplanes

6.4.2.1 GD Method

For a fuel system with integral tanks see 6.4.2.2.

For a fuel system with self-sealing bladder cells:

$$W_{fs} = 41.6\{(W_F/K_{fsp})/100\}^{0.818} + W_{supp} \quad (6.20)$$

For a fuel system with non-self-sealing bladder cells:

$$W_{fs} = 23.1\{(W_F/K_{fsp})/100\}^{0.758} + W_{supp} \quad (6.21)$$

The factor K_{fsp} is defined in 6.4.1.1.

W_{supp} is the weight of the bladder support structure and is given by:

$$W_{supp} = 7.91\{(W_F/K_{fsp})/100\}^{0.854} \quad (6.22)$$

6.4.2.2 Torenbeek Method

For airplanes equipped with non-self-sealing bladder tanks:

$$W_{fs} = 1.6(W_F / K_{fsp})^{0.727} \quad (6.23)$$

For airplanes equipped with integral fuel tanks (wet wing):

$$W_{fs} = 80(N_e + N_t - 1) + 15(N_t)^{0.5}(W_F / K_{fsp})^{0.333} \quad (6.24)$$

A comparison of fuel system weight estimated from Torenbeek and GD methods is shown in Figure 6.3.

6.4.3 Military Patrol, Bomb and Transport Airplanes

For basic fuel system weights, see Sub-Section 6.4.2.

Many military airplanes carry in flight refuelling systems. In addition, many are equipped with fuel dumping systems. The weights of these systems may be estimated from:

For in-flight refuelling:

$$W_{inflref} = 13.64\{(W_F/K_{fsp})/100\}^{0.392} \quad (6.25)$$

For fuel dumping:

$$W_{fd} = 7.38\{(W_F/K_{fsp})/100\}^{0.458} \quad (6.26)$$

6.4.4 Fighter and Attack Airplanes

See Sub-Sections 6.4.2 and 6.4.3.

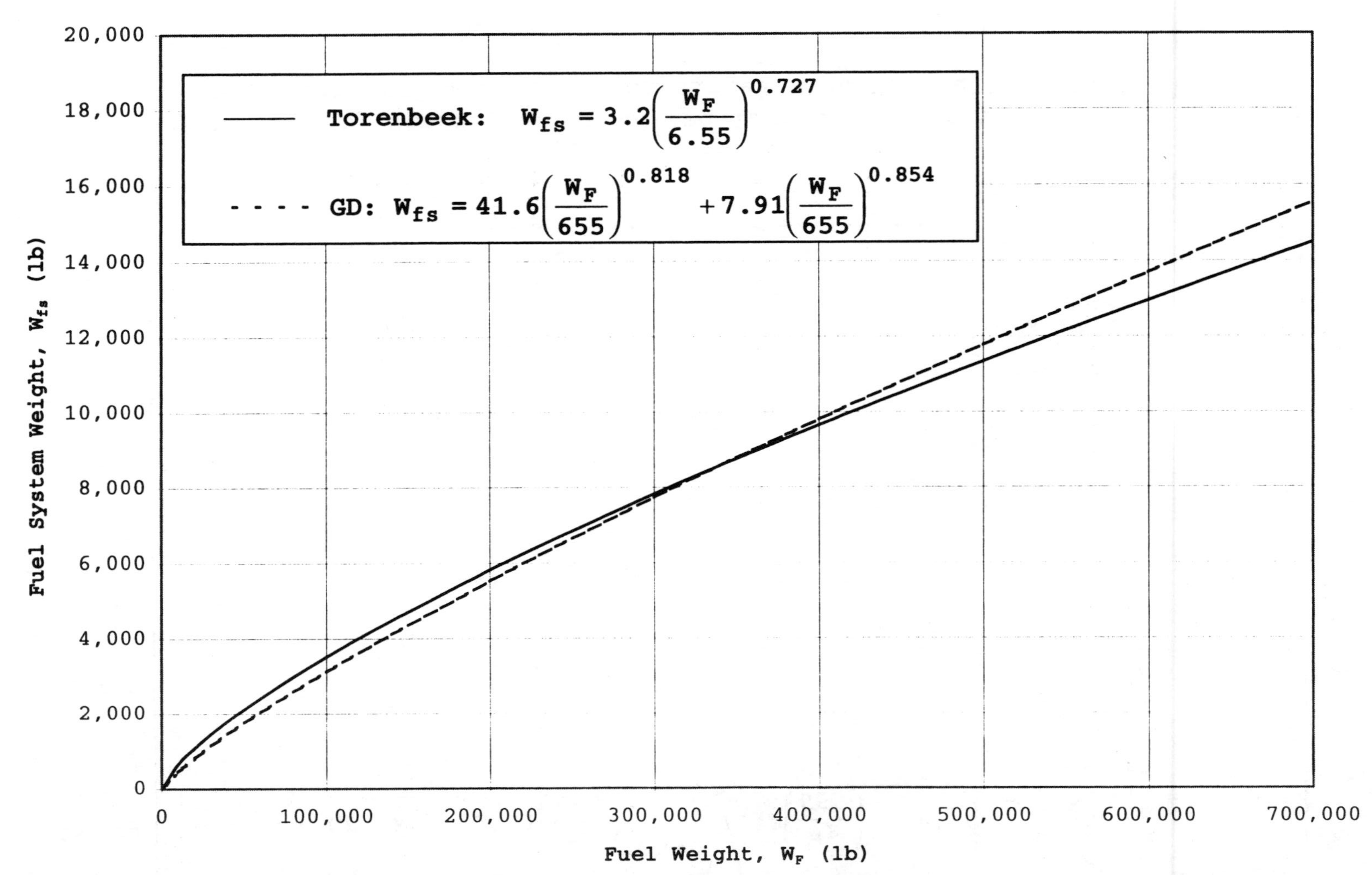

Figure 6.3a: Comparison of GD and Torenbeek Equations for Fuel System Weight Estimation for Airplanes with Self-sealing Tank

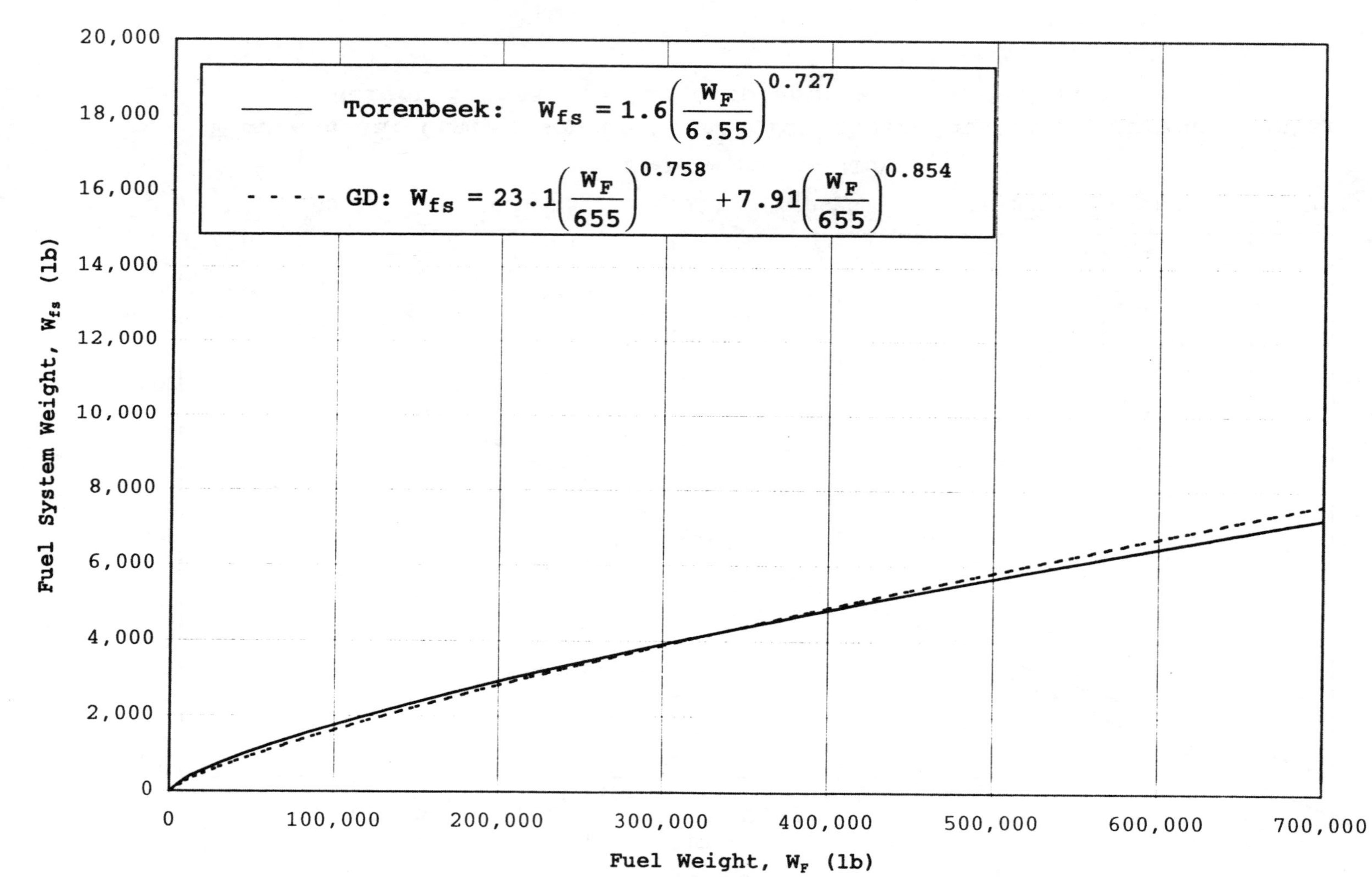

Figure 6.3b: Comparison of GD and Torenbeek Equations for Fuel System Weight Estimation for Airplanes with Non-self-sealing Tank

6.5 PROPULSION SYSTEM WEIGHT ESTIMATION

Depending on airplane type, the propulsion system weight, W_p is either given as a function of total engine weight and/or mission fuel or by:

$$W_p = W_{ec} + W_{ess} + W_{pc} + W_{osc}, \text{ where:} \qquad (6.27)$$

W_{ec} = weight of engine controls in lbs

W_{ess} = weight of engine starting system in lbs

W_{pc} = weight of propeller controls in lbs

W_{osc} = weight of oil system and oil cooler in lbs

6.5.1 General Aviation Airplanes

6.5.1.1 Cessna method

Use actual data.

6.5.1.2 USAF Method

W_p is included in Eqn.(6.3).

6.5.1.3 Torenbeek Method

W_p is included in Eqn.(6.3).

6.5.2 Commercial Transport Airplanes

6.5.2.1 GD Method

Engine controls:

For fuselage/wing-root mounted jet engines:

$$W_{ec} = K_{ec}(l_f N_e)^{0.792} \qquad (6.28)$$

The factor K_{ec} takes on the following values:

K_{ec} = 0.686 for non-afterburning engines
= 1.080 for afterburning engines

For wing mounted jet engines:

$$W_{ec} = 88.46\{(l_f + b)N_e/100\}^{0.294} \qquad (6.29)$$

For wing mounted turboprops:

$$W_{ec} = 56.84\{(l_f + b)N_e/100\}^{0.514} \quad (6.30)$$

For wing mounted piston engines:

$$W_{ec} = 60.27\{(l_f + b)N_e/100\}^{0.724} \quad (6.31)$$

Definition of terms:

N_e = number of engines

l_f = fuselage length in ft

b = wing span in ft

Engine starting systems:

For airplanes with one or two jet engines using cartridge or pneumatic starting systems:

$$W_{ess} = 9.33(W_e/1,000)^{1.078} \quad (6.32)$$

For airplanes with four or more jet engines using pneumatic starting systems:

$$W_{ess} = 49.19(W_e/1,000)^{0.541} \quad (6.33)$$

For airplanes with jet engines using electric starting systems:

$$W_{ess} = 38.93(W_e/1,000)^{0.918} \quad (6.34)$$

For airplanes with turboprop engines using pneumatic starting systems:

$$W_{ess} = 12.05(W_e/1,000)^{1.458} \quad (6.35)$$

For airplanes with piston engines using electric starting systems:

$$W_{ess} = 50.38(W_e/1,000)^{0.459} \quad (6.36)$$

Propeller controls:

For turboprop engines:

$$W_{pc} = 0.322(N_{bl})^{0.589}\{(N_p D_p P_{TO}/N_e)/1,000\}^{1.178} \quad (6.37)$$

For piston engines:

$$W_{pc} = 4.552(N_{bl})^{0.379}\{(N_p D_p P_{TO}/N_e)/1,000\}^{0.759} \quad (6.38)$$

Definition of term:

W_e = total weight of all engines in lbs

6.5.2.2 Torenbeek Method

For airplanes with turbojet or turbofan engines using cartridge or pneumatic starting systems, the weight for accessory drives, powerplant controls, starting and ignition systems is:

$$W_{apsi} = 36N_e(dW_F/dt)_{TO} \quad (6.39a)$$

The take-off fuel flow rate, $(dW_F/dt)_{TO}$ has the dimension of lbs/sec.

For airplanes with turboprop engines this weight is:

$$W_{apsi} = 0.4K_b(N_e)^{0.2}(P_{TO})^{0.8} \quad (6.39b)$$

The factor K_b takes on the following values:

K_b = 1.0 without beta controls
= 1.3 with beta controls

It is usually acceptable to assume that:

$$W_{api} = W_p - W_{osc} \quad (6.40)$$

Definition of new terms:

$(dW_F/dt)_{TO}$ = fuel flow at take-off in lbs/sec

P_{TO} = required take-off power in hp

Thrust reversers for jet engines:

The weight of thrust reversers was already included in the engine weight estimate of Eq.(6.6). To obtain a better estimate of the c.g. effect due to thrust reversers a separate weight estimate is needed:

$$W_{tr} = 0.18W_e \quad (6.41)$$

Water injection system:

Water injection systems are used to increase take-off performance of all types of engines. The installation of such a system is optional.

$$W_{wi} = 8.586W_{wtr}/8.35 \qquad (6.42)$$

W_{wtr} = weight of water carried in lbs

Oil system and oil cooler:

$$W_{osc} = K_{osc}W_e \qquad (6.43)$$

The factor K_{osc} takes on the following values:

K_{osc} = 0.00 for jet engines (weight incl. in W_e)
= 0.07 for turboprop engines
= 0.08 for radial piston engines
= 0.03 for horizontally opposed piston engines

6.5.3 Military Patrol Bomb and Transport Airplanes

See Section 6.5.2.

6.5.4 Fighter and Attack Airplanes

See Section 6.5.2.

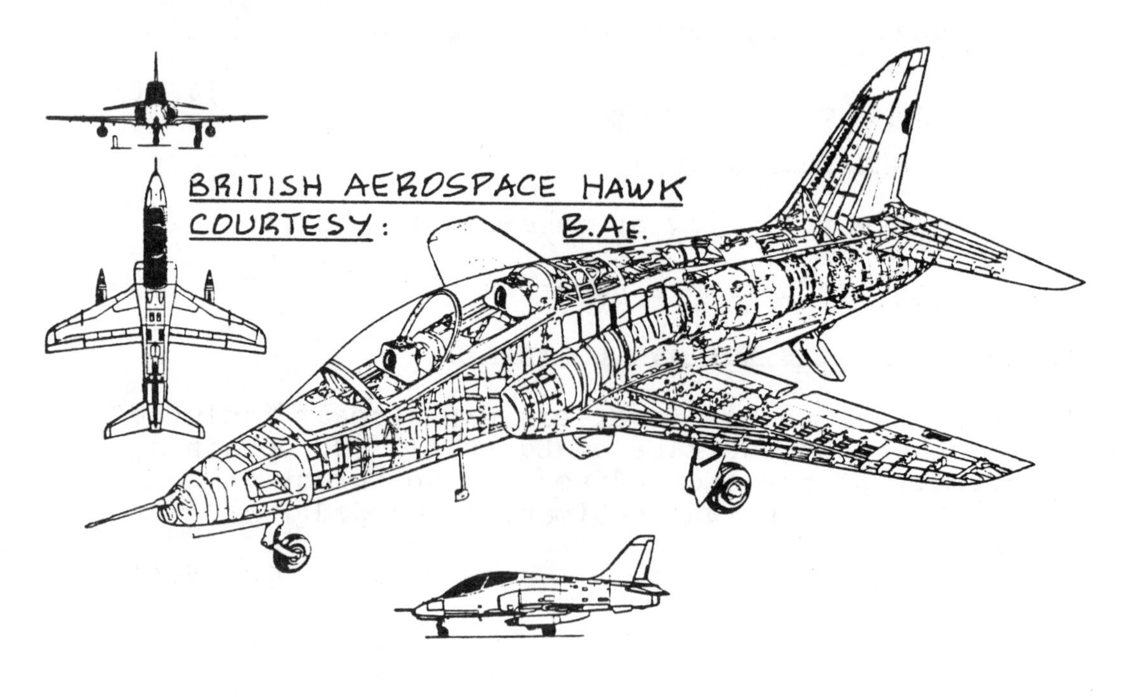

7. CLASS II METHOD FOR ESTIMATING FIXED EQUIPMENT WEIGHT

The list of fixed equipment carried on board airplanes varies significantly with airplane type and airplane mission. In this chapter it will be assumed that the following items are to be included in the fixed equipment category:

7.1. Flight control system, W_{fc}

7.2. Hydraulic and pneumatic System, W_{hps}

7.3. Electrical system, W_{els}

7.4. Instrumentation, avionics and electronics, W_{iae}

7.5. Air-conditioning, pressurization, anti- and de-icing system, W_{api}

7.6. Oxygen system, W_{ox}

7.7. Auxiliary power unit (APU), W_{apu}

7.8. Furnishings, W_{fur}

7.9. Baggage and cargo handling equipment, W_{bc}

7.10 Operational items, W_{ops}

7.11. Armament, W_{arm}

7.12. Guns, launchers and weapons provisions, W_{glw}

7.13. Flight test instrumentation, W_{fti}

7.14. Auxiliary gear, W_{aux}

7.15. Ballast, W_{bal}

7.16. Paint, W_{pt}

7.17. W_{etc}

Therefore:

$$W_{feq} = W_{fc} + W_{hps} + W_{els} + W_{iae} + W_{api} + W_{ox} + W_{apu} + W_{fur} + W_{bc} + W_{ops} + W_{arm} + W_{glw} + W_{fti} + W_{aux} + W_{bal} + W_{pt} + W_{etc} \quad (7.1)$$

The exact definition of which item belongs in a particular fixed equipment category is hard to find. The category W_{etc} was added to cover any items not specifically listed.

Methods for predicting weights of typical fixed equipment items are presented for the following types of airplanes:

1. General Aviation Airplanes
2. Commercial Transport Airplanes
3. Military Patrol, Bomb and Transport Airplanes
4. Fighters and Attack Airplanes

The reader should always consult actual fixed equipment weight data for similar airplanes. Appendix A presents this information for a large number of airplanes.

7.1 FLIGHT CONTROL SYSTEM WEIGHT ESTIMATION

7.1.1 General Aviation Airplanes

7.1.1.1 Cessna Method

$$W_{fc} = 0.0168W_{TO}, \tag{7.2}$$

where: W_{TO} = take-off weight in lbs

This equation applies only to airplanes under 8,000 lbs take-off weight with mechanical flight controls. The equation includes all flight control system hardware: cables, pulleys, pushrods, cockpit controls plus any required back-up structure.

Airplanes in this category all tend to have two sets of flight controls in the cockpit.

7.1.1.2 USAF Method

For airplanes with un-powered flight controls:

$$W_{fc} = 1.066(W_{TO})^{0.626} \tag{7.3}$$

For airplanes with powered flight controls:

$$W_{fc} = 1.08(W_{TO})^{0.7} \tag{7.4}$$

7.1.1.3 Torenbeek Method

For airplanes with un-powered, unduplicated flight controls:

$$W_{fc} = 0.33(W_{TO})^{2/3} \quad (7.5)$$

7.1.2 Commercial Transport Airplanes

7.1.2.1 GD Method

The following equation applies to business jets as well as to commercial transport airplanes:

$$W_{fc} = 56.01 \{(W_{TO})(\bar{q}_D)/100,000\}^{0.576}, \quad (7.6)$$

where: $\bar{q}_D$ is the design dive dynamic pressure in psf

7.1.2.2 Torenbeek Method

$$W_{fc} = K_{fc}(W_{TO})^{2/3} \quad (7.7)$$

The constant K_{fc} takes on the following values:

K_{fc} = 0.44 for airplanes with un-powered flight controls
= 0.64 for airplanes with powered flight controls

If leading edge devices are employed, these estimates should be multiplied by a factor 1.2. If lift dumpers are employed, a factor 1.15 should be used.

7.1.3 Military Patrol, Bomb and Transport Airplanes

7.1.3.1 GD Method

For transport airplanes:

$$W_{fc} = 15.96\{(W_{TO})(\bar{q}_L)/100,000\}^{0.815}, \quad (7.8)$$

where: $\bar{q}_L$ is the design dive dynamic pressure in psf

For Bombers:

$$W_{fc} = 1.049\{(S_{cs})(\bar{q})/1,000\}^{1.21}, \quad (7.9)$$

where: S_{sc} is the total control surface area in ft^2

Note: these estimates include the weight of all associated hydraulic and/or pneumatic systems!

7.1.4 Fighters and Attack Airplanes

7.1.4.1 GD Method

For USAF fighters:

$$W_{fc} = K_{fcf}(W_{TO}/1,000)^{0.581} \qquad (7.10)$$

The constant K_{fcf} takes on the following values:

K_{fcf} = 106 for airplanes with elevon control and no horizontal tail

= 138 for airplanes with a horizontal tail

= 168 for airplanes with a variable sweep wing

For USN fighters and attack airplanes:

$$W_{fc} = 23.77(W_{TO}/1,000)^{1.1} \qquad (7.11)$$

Note: these estimates include the weight of all associated hydraulic and/or pneumatic systems.

Certain airplanes require a center of gravity control system. This is normally implemented using a fuel transfer system. The extra weight due to a c.g. control system may be estimated from:

$$W_{fc_{cg}} = 23.38\{(W_F/K_{fsp})/100\}^{0.442}, \qquad (7.12)$$

where: W_F is the mission fuel weight in lbs

K_{fsp} = 6.55 lbs /gal for JP-4

7.2 HYDRAULIC AND/OR PNEUMATIC SYSTEM WEIGHT ESTIMATION

As seen in Section 7.1 the weight of the hydraulic and/or pneumatic system needed for powered flight controls is usually included in the flight control system weight prediction.

The following weight ratios may be used to determine the hydraulic system weight separately:

For business jets: 0.0070 - 0.0150 of W_{TO}

For regional turboprops: 0.0060 - 0.0120 of W_{TO}

For commercial transports: 0.0060 - 0.0120 of W_{TO}

For military patrol, transport and bombers:

0.0060 - 0.0120 of W_{TO}

For fighters and attack airplanes:

0.0050 - 0.0180 of W_{TO}

The reader should consult the detailed weight data in Appendix A for more precise information.

7.3 ELECTRICAL SYSTEM WEIGHT ESTIMATION

The reader should consult the detailed weight data in Appendix A for electrical system weights of specific airplanes.

7.3.1 General Aviation Airplanes

7.3.1.1 Cessna Method

$$W_{els} = 0.0268W_{TO} \quad (7.13)$$

7.3.1.2 USAF Method

$$W_{els} = 426\{(W_{fs} + W_{iae})/1,000\}^{0.51} \quad (7.14)$$

Note that the electrical system weight in this case is given as a function of the weight of the fuel system plus the weight of instrumentation, avionics and electronics.

7.3.1.3 Torenbeek Method

$$W_{hps} + W_{els} = 0.0078(W_E)^{1.2}, \tag{7.15}$$

where: W_E is the empty weight in lbs

7.3.2 Commercial Transport Airplanes

7.3.2.1 GD Method

$$W_{els} = 1,163\{(W_{fs} + W_{iae})/1,000\}^{0.506} \tag{7.16}$$

7.3.2.2 Torenbeek Method

For propeller driven transports:

$$W_{hps} + W_{els} = 0.325(W_E)^{0.8} \tag{7.17}$$

For jet transports:

$$W_{els} = 10.8(V_{pax})^{0.7}\{1 - 0.018(V_{pax})^{0.35}\}, \tag{7.18}$$

where: V_{pax} is the passenger cabin volume in ft^3

7.3.3 Military Patrol, Bomb and Transport Airplanes

7.3.3.1 GD Method

For transport airplanes:

Use Eqn.(7.15)

For Bombers:

$$W_{els} = 185\{(W_{fs} + W_{iae})/1,000\}^{1.268} \tag{7.19}$$

7.3.4 Fighters and Attack Airplanes

7.3.4.1 GD Method

For USAF fighters:

$$W_{els} = 426\{(W_{fs} + W_{iae})/1,000\}^{0.51} \tag{7.20}$$

For USN fighters and attack airplanes:

$$W_{els} = 347\{(W_{fs} + W_{iae})/1,000\}^{0.509} \tag{7.21}$$

7.4 WEIGHT ESTIMATION FOR INSTRUMENTATION, AVIONICS AND ELECTRONICS

The reader should consult the detailed weight data in Appendix A for weights of instrumentation, avionics and electronics for specific airplanes. Another important source of weight data on actual avionics and electronics systems for civil airplanes is Reference 18. For data on military avionics systems the reader should consult Reference 13, Tables 8-1 and 8-2.

Important comment: The weight equations given in this section are obsolete for modern EFIS type cockpit installations and for modern computer based flight management and navigation systems. The equations provided are probably conservative.

7.4.1 General Aviation Airplanes

7.4.1.1 Torenbeek Method

For single engine propeller driven airplanes:

$$W_{iae} = 33N_{pax}, \tag{7.22}$$

where: N_{pax} is the number of passengers, including the crew

For multi-engine propeller driven airplanes:

$$W_{iae} = 40 + 0.008W_{TO} \tag{7.23}$$

7.4.2 Commercial Transport Airplanes

7.4.2.1 GD Method (Modified)

For the weight of instruments:

$$W_i =$$

$$\underbrace{N_{pil}\{15 + 0.032(W_{TO}/1,000)\}}_{\text{flight instruments}} + \underbrace{N_e\{5 + 0.006(W_{TO}/1,000)\}}_{\text{engine instruments}} +$$

$$\underbrace{+ 0.15(W_{TO}/1,000) + 0.012W_{TO}}_{\text{other instruments}} \tag{7.24}$$

where: N_{pil} is the number of pilots

N_e is the number of engines

7.4.2.2 Torenbeek Method

For regional transports:

$$W_{iae} = 120 + 20N_e + 0.006W_{TO} \quad (7.25)$$

For jet transports:

$$W_{iae} = 0.575(W_E)^{0.556}(R)^{0.25}, \quad (7.26)$$

where: W_E is the empty weight in lbs

R is the maximum range in nautical miles

7.4.3 Military Patrol, Bomb and Transport Airplanes

Use Sub-section 7.4.2.

7.4.4 Fighter and Attack Airplanes

Use Sub-section 7.4.2.

7.5 WEIGHT ESTIMATION FOR AIR-CONDITIONING, PRESSURIZATION, ANTI- AND-DEICING SYSTEMS

7.5.1 General Aviation Airplanes

7.5.1.1 USAF Method

$$W_{api} = 0.265(W_{TO})^{0.52}(N_{pax})^{0.68} \times (W_{iae})^{0.17}(M_D)^{0.08}, \quad (7.27)$$

where: N_{pax} is the number of passengers, including the crew

M_D is the design dive Mach number

7.5.1.2 Torenbeek Method

For single engine, unpressurized airplanes:

$$W_{api} = 2.5N_{pax} \quad (7.28)$$

For multi-engine, unpressurized airplanes:

$$W_{api} = 0.018W_E \quad (7.29)$$

7.5.2 Commercial Transport Airplanes

7.5.2.1 GD Method

For pressurized airplanes:

$$W_{api} = 469\{V_{pax}(N_{cr} + N_{pax})/10,000\}^{0.419} \qquad (7.30)$$

7.5.2.2 Torenbeek Method

For pressurized airplanes:

$$W_{api} = 6.75(l_{pax})^{1.28} \qquad (7.31)$$

where l_{pax} = length of the passenger cabin in ft

7.5.3 Military Patrol, Bomb and Transport Airplanes

7.5.3.1 GD Method

$$W_{api} = K_{api}(V_{pr}/100)^{0.242} \qquad (7.32)$$

The constant K_{api} takes on the following values:

K_{api} = 887 for subsonic airplanes with wing and tail anti-icing

= 610 for subsonic airplanes without anti-icing

= 748 for supersonic airplanes without anti-icing

V_{pr} = pressurized volume in ft^3

7.5.4 Fighters and Attack airplanes

7.5.4.1 GD Method

For low subsonic airplanes:

$$W_{api} = K_{api}\{(W_{iae} + 200N_{cr})/1,000\}^{0.538} \qquad (7.33)$$

The constant K_{api} takes on the following values:

K_{api} = 212 for airplanes with wing and tail anti-icing

= 109 for airplanes without anti-icing

For high subsonic and for supersonic airplanes:

$$W_{api} = 202\{(W_{iae} + 200N_{cr})/1,000\}^{0.735} \qquad (7.34)$$

7.6 WEIGHT ESTIMATION FOR THE OXYGEN SYSTEM

7.6.1 General Aviation Airplanes

Use Sub-section 7.6.2.

7.6.2 Commercial Transport Airplanes

7.6.2.1 GD Method

$$W_{ox} = 7(N_{cr} + N_{pax})^{0.702} \tag{7.35}$$

7.6.2.2 Torenbeek Method

For commercial transport airplanes and for business type airplanes:

For flights below 25,000 ft:

$$W_{ox} = 20 + 0.5N_{pax} \tag{7.36}$$

For short flights above 25,000 ft:

$$W_{ox} = 30 + 1.2N_{pax} \tag{7.37}$$

For extended overwater flights:

$$W_{ox} = 40 + 2.4N_{pax} \tag{7.38}$$

7.6.3 Military Patrol, Bomb and Transport airplanes

Use Sub-section 7.6.2.

7.6.4 Fighters and Attack airplanes

7.6.4.1 GD Method

$$W_{ox} = 16.9(N_{cr})^{1.494} \tag{7.39}$$

7.7 AUXILIARY POWER UNIT WEIGHT ESTIMATION

Auxiliary power units are often used in transport or patrol type airplanes, commercial as well as military.

Actual APU manufacturer data should be used, where possible. Reference 8 contains data on APU systems, under 'Engines'.

From the detailed weight statements in Appendix A it is possible to derive weight fractions for these systems as a function of the take-off weight, W_{TO}. The following ranges are typical of these weight fractions:

$$W_{apu} = (0.004 \text{ to } 0.013)W_{TO} \tag{7.40}$$

7.8 FURNISHINGS WEIGHT ESTIMATION

The furnishings category normally includes the following items:

1. Seats, insulation, trim panels, sound proofing, instrument panels, control stands, lighting and wiring
2. Galley (pantry) structure and provisions
3. Lavatory (toilet) and associated systems
4. Overhead luggage containers, hatracks, wardrobes
5. Escape provisions, fire fighting equipment

Note: the associated consumable items such as potable water, food, beverages and toilet chemicals and papers are normally included in a weight category referred to as: Operational Items: W_{ops}, see Section 7.10.

The reader is referred to the detail weight statements in Appendix A for actual furnishings weight data on specific airplanes.

7.8.1 General Aviation airplanes

7.8.1.1 Cessna Method

$$W_{fur} = 0.412(N_{pax})^{1.145}(W_{TO})^{0.489}, \tag{7.41}$$

where: N_{pax} is the number of passengers including the crew

7.8.1.2 Torenbeek Method:

For single engine airplanes:

$$W_{fur} = 5 + 13N_{pax} + 25N_{row}, \qquad (7.42)$$

where: N_{row} is the number of seat rows

For multi engine airplanes:

$$W_{fur} = 15N_{pax} + 1.0V_{pax+cargo}, \qquad (7.43)$$

where: $V_{pax+cargo}$ is the volume of the passenger cabin plus the cargo volume in ft^3

7.8.2 Commercial Transport Airplanes

The weight of furnishings varies considerably with airplane type and with airplane mission. This weight item is a considerable fraction of the take-off weight of most airplanes, as the data in Appendix A illustrate.

Reference 14 contains a very detailed method for estimating the furnishings weight for commercial transport airplanes.

7.8.2.1 GD Method

$$W_{fur} = \qquad (7.44)$$

$$55N_{fdc} + 32N_{pax} + 15N_{cc} + K_{lav}(N_{pax})^{1.33} + K_{buf}(N_{pax})^{1.12}$$

fdc sts pax sts cc sts lavs + water food prov.

$$+ 109\{N_{pax}(1 + P_c)/100\}^{0.505} + 0.771(W_{TO}/1,000)$$

cabin windows miscellaneous

The factor K_{lav} takes on the following values:

K_{lav} = 3.90 for business airplanes
= 0.31 for short range airplanes
= 1.11 for long range airplanes

The factor K_{buf} takes on the following values:

K_{buf} = 1.02 for short ranges
= 5.68 for very long ranges

The term P_c is the design ultimate cabin pressure in psi. The value of P_c depends on the design altitude for the pressure cabin.

7.8.2.2 Torenbeek Method

$$W_{fur} = 0.211(W_{TO} - W_F)^{0.91} \quad (7.45)$$

In commercial transports it is usually desirable to make more detailed estimates than possible with Eqn.(7.45). Particularly if a more accurate location of the c.g. of items which contribute to the furnishings weight is needed, a more detailed method may be needed. Reference 14 contains the necessary detailed information.

7.8.3 Military Patrol, Bomb and Transport Airplanes

7.8.3.1 GD Method

$$W_{fur} = \text{Sum} \downarrow \text{ in the tabulation below.} \quad (7.46)$$

Type	Patrol	Bomb	Transport
Crew Ej. Seats	$K_{st}(N_{cr})^{1.2}$	$K_{st}(N_{cr})^{1.2}$	
	K_{st} = 0 for no ejection seat = 149 with survival kit = 100 without survival kit		
Crew Seats	$83(N_{cr})^{0.726}$	same	same
Passenger Seats			$32(N_{pax})$
Troop Seats			$11.2(N_{troop})$
Lav. and Water			$1.11(N_{pax})^{1.33}$
Misc.	$0.0019(W_{TO})^{0.839}$		$0.771(W_{TO}/1{,}000)$

7.8.4 Fighters and Attack Airplanes

$$W_{fur} = \quad (7.47)$$

$$= 22.9(N_{cr}\bar{q}_D/100)^{0.743} + 107(N_{cr}W_{TO}/100{,}000)^{0.585}$$

ejection seats   Misc. and emergency eqpmt

7.9 WEIGHT ESTIMATION OF BAGGAGE AND CARGO HANDLING EQUIPMENT

The GD method gives for military passenger transports:

$$W_{bc} = K_{bc}(N_{pax})^{1.456} \quad (7.48)$$

The constant K_{bc} takes on the following values:

K_{bc} = 0.0646 without preload provisions
= 0.316 with preload provisions

The Torenbeek method gives for commercial cargo airplanes:

$$W_{bc} = 3S_{ff}, \quad (7.49)$$

where: S_{ff} is the freight floor area in ft^2.

For baggage and for cargo containers, the following weight estimates may be used:

freight pallets:	88x108 in	225 lbs
(including nets)	88x125 in	262 lbs
	96x125 in	285 lbs

containers: 1.6 lbs/ft^3 (For container dimensions, see Part III.)

7.10 WEIGHT ESTIMATION OF OPERATIONAL ITEMS

Typical weights counted in operational items are:

*Food *Potable water *Drinks

*China *Lavatory supplies

Observe that Eqn. (7.44) includes these operational items. For more detailed information on operational items the reader should consult Reference 14, p.292.

7.11 ARMAMENT WEIGHT ESTIMATION

The category armament can contain a wide variety of weapons related items as well as protective shielding for the crew. Typical armament items are:

*Firing systems *Fire control systems

*Bomb bay or missile doors *Armor plating

*Weapons ejection systems

Note that the weapons themselves as well as any ammunition are not normally included in this item.

Appendix A contains data on 'armament' weight for several types of military airplanes.

7.12 WEIGHT ESTIMATION FOR GUNS, LAUNCHERS AND WEAPONS PROVISIONS

For detailed data on guns, launchers and other military weapons provisions the reader is referred to Part IV, Chapter 3.

Note: Ammunition, bombs, missiles, and most types of external stores are normally counted as part of the payload weight, W_{PL} in military airplanes.

7.13 WEIGHT ESTIMATION OF FLIGHT TEST INSTRUMENTATION

During the certification phase of most airplanes a significant amount of flight test instrumentation and associated hardware is carried on board. The magnitude of W_{fti} depends on the type of airplane and the types of flight tests to be performed. Appendix A contains weight data for flight test instrumentation carried on a number of NASA experimental airplanes (Tables A13.1-A13.4).

7.14 WEIGHT ESTIMATION FOR AUXILIARY GEAR

This item encompasses such equipment as:

*fire axes *sextants *unaccounted items

An item referred to as 'manufacturers variation' is sometimes included in this category as well. A safe assumption is to set:

$$W_{aux} = 0.01W_E \qquad (7.50)$$

7.15 BALLAST WEIGHT ESTIMATION

When looking over the weight statements for various airplanes in Appendix A, the reader will make the startling discovery that some airplanes carry a significant amount of ballast. This can have detrimental effects on speed, payload and range performance.

The following reasons can be given for the need to include ballast in an airplane:

1. The designer 'goofed' in the weight and balance calculations

2. To achieve certain aerodynamic advantages it was judged necessary to locate the wing or to size the empennage so that the static margin became insufficient. This problem can be solved with ballast. In this case, carrying ballast may in fact turn out to be advantageous.

3. To achieve flutter stability within the flight envelope ballast weights are sometimes attached to the wing and/or to the empennage.

Note: balance weights associated with flight control surfaces are not counted as ballast weight.

The amount of ballast weight required is determined with the help of the X-plot. Construction and use of the X-plot is discussed in Part II, Chapter 11. The Class II weight and balance method discussed in Chapter 9 of this part may also be helpful in determining the amount of ballast weight required to achieve a certain amount of static margin.

7.16 ESTIMATING WEIGHT OF PAINT

Transport jets and camouflaged military airplanes carry a considerable amount of paint. The amount of paint weight is obviously a function of the extent of surface coverage. For a well painted airplane a reasonable estimate for the weight of paint is:

$$W_{pt} = 0.003W_{TO} \text{ to } 0.006W_{TO} \qquad (7.51)$$

7.17 ESTIMATING WEIGHT OF W_{etc}

This weight item has been included to cover any items which do not normally fit in any of the previous weight categories.

8. LOCATING COMPONENT CENTERS OF GRAVITY

The purpose of this chapter is to provide guidelines for the determination of the location of centers of gravity for individual airplane components. Knowledge of component c.g. locations is essential in both Class I and Class II weight and balance analyses as discussed in Chapter 10 of Part II and Chapter 4 of this book.

In Part II, Chapter 10, Table 10.2 provides a summary of c.g. locations for the major structural components of the airplane only. In this chapter a slightly more extensive data base is provided. The presentation of component c.g. locations follows the weight breakdowns of Chapters 5-7:

8.1 C.G. Locations of Structural Components
8.2 C.G. Locations of Powerplant Components
8.3 C.G. Locations for Fixed Equipment

8.1 C.G. LOCATIONS OF STRUCTURAL COMPONENTS

Table 8.1 lists the most likely c.g. locations for major structural components. There is no substitute for common sense: if the preliminary structural arrangement of Part III (Step 19 of p.d. sequence 2, Part II) suggests that a given structural component has a different mass distribution than is commonly the case, an 'educated guess' must be made as to the effect on the c.g. of that component.

Example: Looking at the threeview of the GP-180 of Figure 3.47, p.86, Part II it is obvious that there is a concentration of primary structure at the aft end of the fuselage. The fuselage c.g. should therefore not be placed at 38-40 percent of the fuselage length, but probably at 55 to 60 percent.

8.2 C.G. LOCATIONS OF POWERPLANT COMPONENTS

Table 8.2 lists the most likely c.g. locations for powerplant components. Note that for engine c.g. locations manufacturers data should be used. 'Guessing' at engine c.g. locations is not recommended!

8.3 C.G. LOCATIONS OF FIXED EQUIPMENT

Table 8.3 lists guidelines for locating centers of gravity of fixed equipment components.

Table 8.1 Center of Gravity Location of Structural Components

Component:	Center of gravity location:
Wing (half):	Unswept wing: 38-42 percent chord from the L.E. at 40 percent of the semi-span. Swept wing: 70 percent of the distance between the front and rear spar behind the front spar at 35 percent of the semi-span
Horizontal Tail: (half)	Regardless of sweep angle: 42 percent chord from the L.E. at 38 percent of the semi-span.
Vertical Tail: (low tail)	Regardless of sweep angle: 42 percent chord from the L.E. at 38 percent vertical tail span from the root chord.
Vertical Tail: (T-tail) root	Regardless of sweep angle: 42 percent chord from the L.E. at 55 percent vertical tail span from the root chord.
Vertical Tail: (cruciform)	Regardless of sweep angle: 42 percent chord from the L.E. at between 38 and 55 percent vertical tail span from the root chord. Interpolate according to z_h/b_v.
Fuselage: Caution: Do not count the propeller spinner in fuselage or nacelle length!	Distances are given as a fraction of the fuselage length: Single engine tractors: 0.32-0.35 Single engine pusher: 0.45-0.48 Propeller driven twins: 0.38-0.40 (tractors on wing) Propeller driven twins: 0.50-0.53 (pushers on wing) Jet transports: 0.42-0.45 (wing mounted engines) Jet transports: 0.47-0.50 (rear fuselage mounted engines) Fighters: 0.45 (engines buried in the fuselage)
Tail booms:	0.40-0.45 of boom length starting from most forward structural attachment of the boom.
Nacelles:	0.40 of nacelle length from nacelle nose
Landing gear:	at 0.50 of strutlength for gears with mostly vertical struts

Table 8.2 Center of Gravity Location of Powerplant Components

Component:	Center of Gravity Location:
Engine(s)	Use manufacturers data
Air induction system	Use the c.g. of the gross shell area of the inlets
Propellers	On the spin axis, in the propeller spin plane
Fuel system	Refer to the fuel system layout diagram required as part of Step 17 in p.d. sequence II, Part II, p.18.
Filled fuel tank	Assuming a prismoidal shape (See figure left), the c.g. is located relative to plane S_1 at:

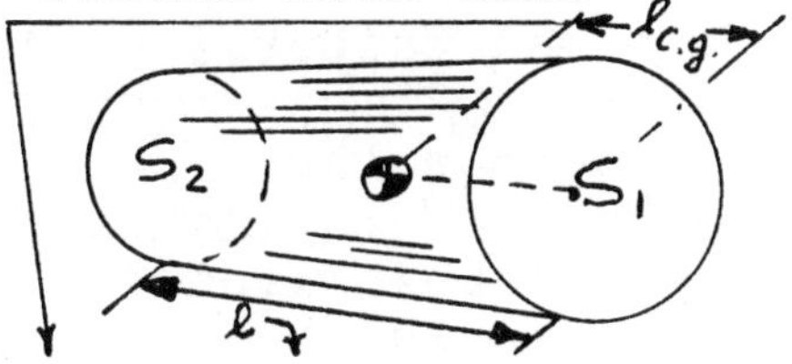

$$l_{cg} = (1/4)\{S_1 + 3S_2 + 2(S_1S_2)^{1/2}\}/\{S_1 + S_2 + (S_1S_2)^{1/2}\} \qquad (8.1)$$

Component:	Center of Gravity Location:
Trapped fuel and oil	Trapped fuel is normally located at the bottom of fuel tanks and fuel lines. Trapped oil is normally located close to the engine case.
Propulsion system	Make a list of which items contribute to the propulsion system weight and 'guestimate' their c.g. location by referring to the powerplant installation drawing required in Step 5.10, pages 133 and 134 in Part II.

Table 8.3 Center of Gravity Location of Fixed Equipment

Component:	Center of Gravity Location:
Flight Control System Hydraulic and Pneumatic System Electrical System Instrumentation, Avionics and Electronics Air-conditioning, Pressurization, Anti-icing and de-icing System Oxygen System	Note: for all systems, the c.g. location can be most closely 'guestimated' by referring to the system lay-out diagrams described in Part IV of this text. These system lay-outs were required as part of Step 17 in p.d. sequence 2, Part II, p.18.
Auxiliary Power Unit	See engine manufacturer data.
Furnishings Baggage and Cargo Handling Equipment	Refer to the fuselage internal arrangement drawing required by Steps 4.1 and 4.2 in Part II, pp 107 and 108.
Operational items	See furnishings
Armament	This item is normally close to the cockpit
Guns, launchers and weapons provisions	From manufacturer data.
Flight test instrumentation	A sketch depicting the locations of sensors, recorders operating systems should help in locating the overall c.g. of this item.
Auxiliary gear	Make a list of items in this category and 'guestimate' their c.g. locations.
Ballast	Ballast weights are normally made from lead. Ballast c.g. is thus easily located.
Paint	Centroid of painted areas.

9. CLASS II WEIGHT AND BALANCE ANALYSIS

The basic method used in performing a Class II weight and balance analysis is identical to that used for the Class I weight and balance analysis. The latter was discussed in detail in Part II, Chapter 10. The only difference is, that a more detailed weight statement is used: the Class II weight prediction method of Chapters 4-7 is used.

During this stage of the preliminary design frequent questions which are raised, are:

1. How much does the overall airplane c.g. move as a result of moving some component?

2. How much does the airplane static margin change as a result of moving the wing?

These questions are answered in Sections 9.1 and 9.2 respectively.

9.1 EFFECT OF MOVING COMPONENTS ON OVERALL AIRPLANE CENTER OF GRAVITY

Figure 9.1 illustrates an airplane, its c.g. location and the c.g. location of component i. The overall center of gravity of the airplane is found from:

$$x_{cg} = \left(\sum_{i=1}^{i=n} W_i x_i\right) / \left(\sum W_i\right) \tag{9.1}$$

Evidently:

$$\sum_{i-1}^{i=n} W_i = W \tag{9.2}$$

The rate at which overall airplane c.g. moves, when a component i is moved, can be found by differentiation of Eqn.(9.1):

$$\partial x_{cg} / \partial x_i = (W_i) / \left(\sum_{i=1}^{i=n} W_i\right) \tag{9.3}$$

If component i is moved over a distance Δx_i, the

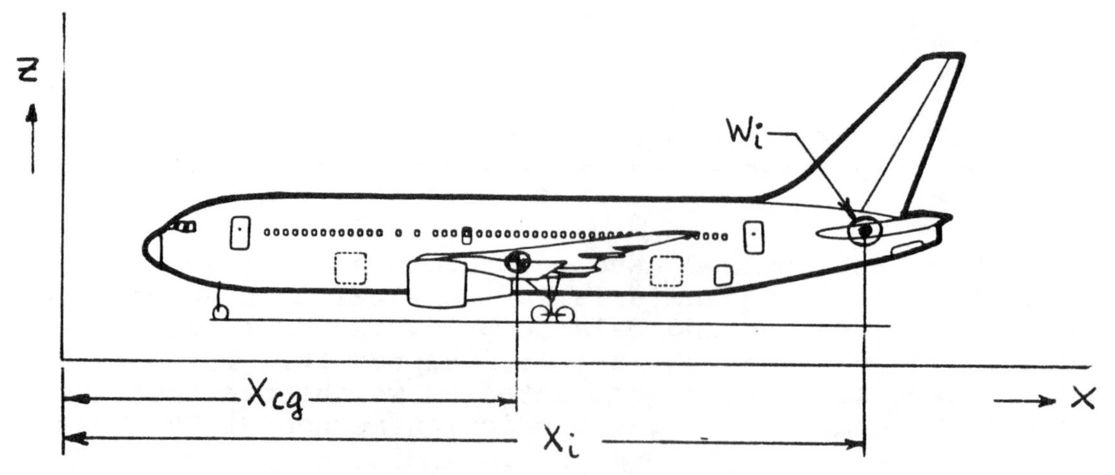

Figure 9.1 Definition of Overall C.G. Location and of Component C.G. Location

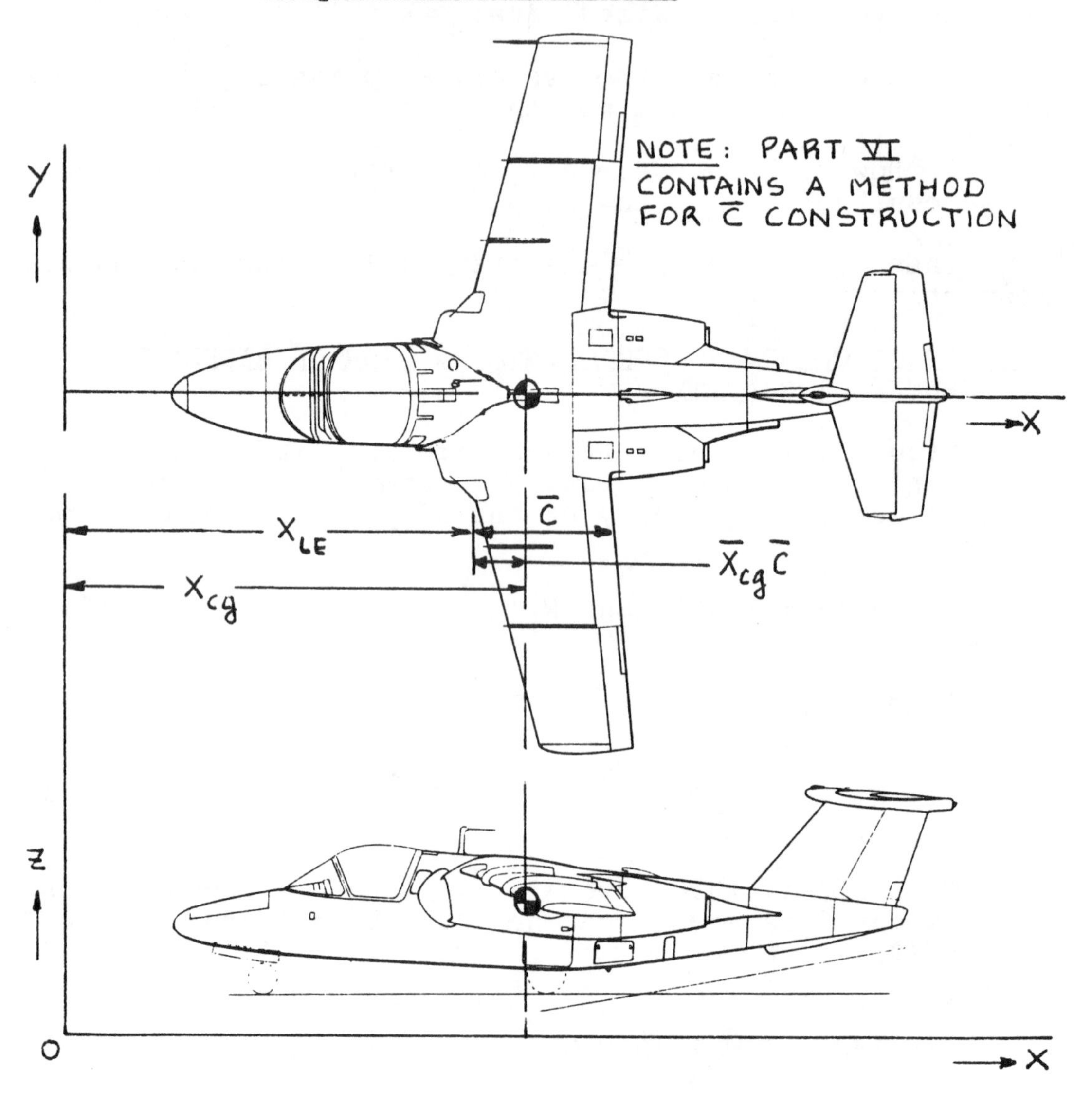

Figure 9.2 Definition of Mean Geometric Chord ($\bar{c}$) Location and Overall C.G. Location

overall airplane c.g. moves over a distance given by:

$$\Delta x_{cg} = (\Delta x_i)(W_i)/(\text{Sum}_{i=1}^{i=n} W_i) \quad (9.4)$$

Equation (9.4) suggests that to move the overall c.g. of the airplane significantly, either a heavy weight component can be moved a small distance or a light weight component can be moved a large distance.

Items which are frequently moved about to achieve satisfactory weight and balance results are: batteries, air-conditioner units, certain 'black' boxes and sometimes just plain ballast. The reader will note from the detailed weight statements in Appendix A that several airplanes carry a relatively large amount of ballast.

9.2 EFFECT OF MOVING THE WING ON OVERALL AIRPLANE CENTER OF GRAVITY AND ON OVERALL AIRPLANE AERODYNAMIC CENTER

Figure 9.2 illustrates an airplane, its mean geometric chord location and its overall c.g. location.

If the leading edge of the mgc of the wing is at fuselage station (FS) x_{LE}, the airplane c.g. in terms of the wing mgc can be written as:

$$\bar{x}_{cg} = (x_{cg} - x_{LE})/\bar{c} \quad (9.5)$$

When a component i is moved over a distance Δx_i, the overall airplane c.g. moves relative to the wing mgc as:

$$\Delta \bar{x}_{cg} = (\Delta x_i)(W_i)/\bar{c}(\text{Sum}_{i=1}^{i=n} W_i) \quad (9.6)$$

For a conventional, tail-aft airplane, its aerodynamic center location can be written as:

$$\bar{x}_{ac} = \{C_1 + C_2(\bar{x}_{ac_h})\}/(1 + C_2) \quad (9.7)$$

$$\text{where: } C_1 = \bar{x}_{ac_w} + \Delta\bar{x}_{ac_{wb}} \quad (9.8)$$

$$C_2 = (C_{L_{\alpha_h}}/C_{L_{\alpha_{wb}}})(1 - d\varepsilon/d\alpha)(S_h/S) \quad (9.9)$$

A detailed derivation of Eqn. (9.7) may be found in Reference 19, p.133.

Part VI contains methods for computing the liftcurve slopes and aerodynamic centers which appear in C_1 and in C_2.

<u>Warning:</u> the wing-body aerodynamic center shift, $\Delta\bar{x}_{ac_{wb}}$ in Eqn. (9.8) is always a negative number: it shifts the a.c. forward!

If the wing is moved aft over a distance Δx_w, the overall airplane c.g. is:

$$\bar{x}_{cg_{new}} = \bar{x}_{cg_{old}} + (\Delta x_w W_w)/(\bar{c}(\mathrm{Sum}_{i=1}^{i=n} W_i) \qquad (9.10)$$

The new a.c. location can now be written as:

$$\bar{x}_{ac_{new}} = \{C_1 + C_2(\bar{x}_{ac_h} - \Delta x_w/\bar{c})\}/(1 + C_2) \qquad (9.11)$$

Equations (9.10) and (9.11) can be used to 'redo' the X-plot of Part II, Chapter 11. This 'redone' X-plot in turn is used to:

1. determine how much the horizontal tail area must be changed as a result of moving the wing

or:

2. determine which other weight components need to be moved and by how much, to maintain some desired level of stability (or instability as the case may be).

For a canard airplane and/or for a three surface airplane similar equations are easily derived. The reader should consult Equations (11.1) and (11.2) in Part II for guidance.

10. CLASS II METHOD FOR ESTIMATING AIRPLANE INERTIAS

The purpose of this chapter is to provide an outline for a Class II method for estimating moments and products of inertia. It will be assumed that the Class II weight estimating method of Chapter 4 has been applied: a rather detailed weight and c.g. breakdown for the airplane is therefore presumed to be available.

The following equations are a slight modification of the general inertia equations 2.22a through 2.22c in Reference 19.

$$I_{xx} = \sum_{i=1}^{i=n} m_i\{(y_i - y_{cg})^2 + (z_i - z_{cg})^2\} \qquad (10.1)$$

$$I_{yy} = \sum_{i=1}^{i=n} m_i\{(z_i - z_{cg})^2 + (x_i - x_{cg})^2\} \qquad (10.2)$$

$$I_{zz} = \sum_{i=1}^{i=n} m_i\{(x_i - x_{cg})^2 + (y_i - y_{cg})^2\} \qquad (10.3)$$

$$I_{xy} = \sum_{i=1}^{i=n} m_i(x_i - x_{cg})(y_i - y_{cg}) \qquad (10.4)$$

$$I_{yz} = \sum_{i=1}^{i=n} m_i(y_i - y_{cg})(z_i - z_{cg}) \qquad (10.5)$$

$$I_{zx} = \sum_{i=1}^{i=n} m_i(z_i - z_{cg})(x_i - x_{cg}) \qquad (10.6)$$

Figure 10.1 defines the coordinates used in these equations.

The reader should recall that for a symmetrical airplane the inertia products I_{xy} and I_{yz} are zero.

Equations (10.1) through (10.6) are valid whenever the weight breakdown contains a 'sufficiently' large number of parts so that the inertia moment and/or product of each part about its own c.g. location is negligible.

Whenever the latter assumption is not satisfied,

equations (10.1) through (10.6) should be all modified as follows:

$$I_{xx} = \sum_{i=1}^{i=n} I_{xx_i} + \sum_{i=1}^{i=n} m_i\{(y_i - y_{cg})^2 + (z_i - z_{cg})^2\} \quad (10.7)$$

The first term in Eqn.(10.7) represents the moment (or product) of inertia of component i about its own center of gravity.

Moments (and products) of inertia of airplane components about their own center of gravity can be computed in a relatively straightforward manner by assuming uniform mass distributions for structural components and by using the 'lumped mass' assumption for distributed systems. An example of the latter would be the airplane fuel system. Major fuel system components such as pumps, bladders and the like can be considered to be concentrated masses distributed around the fuel system c.g.. Equations (10.1) through (10.2) are then used to compute the moments of inertia of the fuel system about its own c.g.

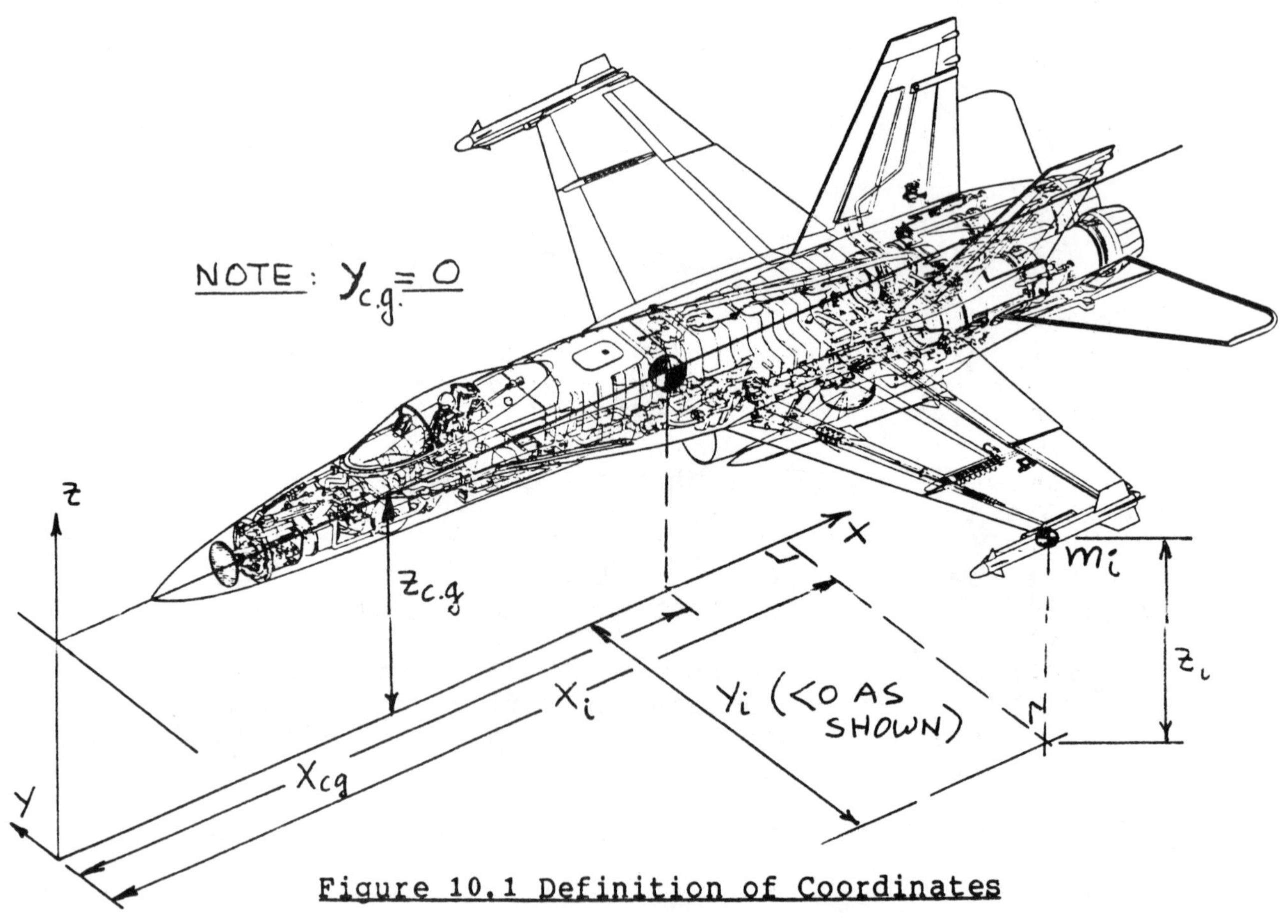

Figure 10.1 Definition of Coordinates

11. REFERENCES

1. Roskam, J., Airplane Design: Part I, Preliminary Sizing of Airplanes.

2. Roskam, J., Airplane Design: Part II, Preliminary Configuration Design and Integration of the Propulsion System.

3. Roskam, J., Airplane Design: Part III, Layout Design of Cockpit, Fuselage, Wing and Empennage: Cutaways and Inboard Profiles.

4. Roskam, J., Airplane Design: Part IV, Layout Design of Landing Gear and Systems.

5. Roskam, J., Airplane Design: Part VI, Preliminary Calculation of Aerodynamic, Thrust and Power Characteristics.

6. Roskam, J., Airplane Design: Part VII, Determination of Stability, Control and Performance Characteristics: FAR and Military Requirements.

7. Roskam, J., Airplane Design: Part VIII, Airplane Cost Estimation and Optimization: Design, Development Manufacturing and Operating.

Note: These books are all published by: Roskam Aviation and Engineering Corporation, Rt4, Box 274, Ottawa, Kansas, 66067, Tel. 913-2421624.

8. Taylor, J.W.R., Jane's All The World Aircraft, Published Annually by: Jane's Publishing Company, 238 City Road, London EC1V 2PU, England. (Issues used: 1945/46, 1968/84)

9. Chawla, J.P., Empirical Formulae for Radii of Gyration of Aircraft, SAWE Paper No.78, Hughes Aircraft Company, Culver City, California, 1952.

10. Anon., Empirical Formulae for Moments of Inertia of Aircraft, Royal Aircraft Establishment, Structures Report No. 28, Farnborough, England, 1948.

11. Garcia, D., Empirical Formulae for Radii of Gyration of Aircraft, Revision A, SAWE Paper No.78A, Republic Aviation Corporation, Farmingdale, Long Island, NY, 1962.

12. Schmitt, R.L., Foreman, K.C., Gertsen, W.M. and Johnson, P.H., Weight Estimation Handbook for Light Aircraft, Cessna Aircraft Company, 1959.

13. Nicolai, L.M., Fundamentals of Aircraft Design, METS, Inc., 6520 Kingsland Court, CA, 95120.

14. Torenbeek, E., Synthesis of Subsonic Airplane Design, Kluwer Boston inc., Hingham, Maine, 1982.

15. Anon., Federal Aviation Regulation, Part 23, Department of Transportation, Federal Aviation Administration, Distribution Requirements Section, M-482.2, Washington D.C., 20590.

16. Anon., Federal Aviation Regulation, Part 25, see Ref. 15.

17. MIL-A-8861(ASG), Military Specification, Airplane Strength and Rigidity, Flight Loads, May 1960.

18. Business and Commercial Aviation (Monthly magazine), 1985 Planning and Purchasing Handbook, April 1985.

19. Roskam, J., Airplane Flight Dynamics and Automatic Flight Controls, Part I, Roskam Aviation and Engineering Corporation, Rt 4, Box 274, Ottawa, Kansas, 66067.

APPENDIX A: DATA SOURCE FOR AIRPLANE COMPONENT WEIGHTS AND FOR WEIGHT FRACTIONS

Tables A1 through A13 present component weight data for the following types of airplanes:

1. Homebuilt propeller driven airplanes: Tables A1.1.

2. Single engine propeller driven airplanes: Tables A2.1 and A2.2.

3. Twin engine propeller driven airplanes: Tables A3.1 and A3.2.

4. Agricultural airplanes: Tables A4.1. At the time of printing no data were available. The reader should use Tables A2 and add spray equipment weights.

5. Business Jets: Tables A5.1 and A5.2.

6. Regional turbopropeller airplanes: Tables A6.1 through A6.3.
   Regional piston/propeller airplanes: Tables A6.4.

7. Jet transports: Tables A7.1 through A7.5.
   Turbopropeller driven transports: Tables A7.6.

8. Military trainers: Tables A8.1.

9. Fighters: Tables A9.1 through A9.5.

10. Military jet transports: Tables A.10.1.
    Military turbopropeller driven transports: Tables A10.2.
    Military piston/propeller driven transports: Table A10.3 and A10.4.
    Military patrol airplanes: Tables A10.5.

11. Flying boats, amphibious and float airplanes: Tables A11.1. At the time of printing no data were available. The reader should use suitable tables in categories 1-10 and account for hull weight with the method of Chapter 4.

12. Supersonic cruise airplanes: Tables A12.1.

13. NACA and NASA X (experimental) airplanes: Tables A13.1 through A13.4.
    CAUTION: most of the X airplanes were built for experimental purposes only. They should not be regarded as 'optimized' for a given mission.

Table A1.1a Group Weight Data for Homebuilt Propeller Driven Airplanes

Type Weight Item, lbs	Bede BD5B	At the time of printing no other data were available
Wing Group	87	
Empennage Group	17	
Fuselage Group	89	
Nacelle Group	0	
Landing Gear Group	32	
Nose Gear	10	
Main Gear	22	
Structure Total	225	
Engine	146	
Air Induct. System	0	
Fuel System	25	
Propeller Install.	5	
Thrust Attenuator	3	
Engine Install.	10	
Power Plant Total	189	
Avionics + Instrum.	15	
Surface Controls	0	
Electrical System	10	
Electronics	0	
Ballast	30	
Parachute	20	
Furnishings + paint	50	
Auxiliary Gear	0	
Fixed Equipm't Total	125	
$W_{oil} + W_{tof}$	2	
Fuel	340	
Payload(pilot)	170	

Table A1.1b Group Weight Data for Homebuilt Propeller Driven Airplanes

Type	Bede BD5B	At the time of printing no other data were available
Flight Design Gross Weight, GW, lbs	1,051	
Structure/GW	0.214	
Power Plant/GW	0.180	
Fixed Equipm't/GW	0.119	
Empty Weight/GW	0.513	
Wing Group/GW	0.083	
Empenn. Group/GW	0.016	
Fuselage Group/GW	0.085	
Nacelle Group/GW	0.000	
Land. Gear Group/GW	0.030	
Take-off Gross Wht, W_{TO}, lbs	1,051	
Empty Weight, W_E, lbs	539	
Wing Group/S, psf	1.8	
Emp. Grp/S_{emp}, psf	1.1	
Ultimate Load Factor, g's	5.7 assumed	
Surface Areas, ft^2		
Wing, S	47.4	
Horiz. Tail, S_h	10.5	
Vert. Tail, S_v	5.0	
Empenn. Area, S_{emp}	15.5	

Table A2.1a Group Weight Data for Single Engine Propeller Driven Airplanes

Type	Cessna 150	172	175	180 **	182	L-19A* **
Weight Item, lbs						
Wing Group	216	226	227	235	235	238
Empennage Group	36	57	57	62	62	64
Fuselage Group	231	353	351	404	400	216
Nacelle Group	22	27	30	32	34	33
Landing Gear Group	104	111	111	112	132	135
Nose Gear						
Main Gear						
Structure Total	609	774	776	845	863	686
Engine	197	254	318	417	417	399
Air Induct. System	2	1	3	1	1	4
Fuel System	17	21	26	26	26	39
Propeller Install.	22	33	33	64	64	46
Engine Install.	28	36	36	37	37	62
Power Plant Total	267	345	416	545	545	550
Avionics + Instrum.	3	4	4	6	6	36
Surface Controls	31	31	31	36	36	47
Electrical System	34	38	38	43	43	86
Electronics	0	0	0	0	0	39
Air Cond. System Anti-icing System	1	1	1	1	1	9
Furnishings	33	85	85	87	87	65
Auxiliary Gear	0	0	0	0	0	3
Fixed Equipm't Total	102	159	159	173	173	285
$W_{oil} + W_{tof}$	11	15	19	22	22	19
Fuel	156	252	312	390	390	252
Payload	398	702	719	734	715	321

*Military observation airplane
**Taildragger

Table A2.1b Group Weight Data for Single Engine Propeller Driven Airplanes

Type	Cessna 150	172	175	180 **	182	L-19A* **
Flight Design Gross Weight, GW, lbs	1,500	2,200	2,350	2,650	2,650	2,100
Structure/GW	0.406	0.352	0.330	0.319	0.326	0.327
Power Plant/GW	0.178	0.157	0.177	0.206	0.206	0.262
Fixed Equipm't/GW	0.068	0.072	0.068	0.065	0.065	0.136
Empty Weight/GW	0.631	0.565	0.561	0.576	0.583	0.727
Wing Group/GW	0.144	0.103	0.097	0.089	0.089	0.113
Empenn. Group/GW	0.024	0.026	0.024	0.023	0.023	0.030
Fuselage Group/GW	0.154	0.160	0.149	0.152	0.151	0.103
Nacelle Group/GW	0.015	0.012	0.013	0.012	0.013	0.016
Land. Gear Group/GW	0.069	0.050	0.047	0.042	0.050	0.064
Take-off Gross Wht, W_{TO}, lbs	1,500	2,200	2,350	2,650	2,650	2,100
Empty Weight, W_E, lbs	946	1,243	1,319	1,526	1,545	1,527
Wing Group/S, psf	1.4	1.4	1.3	1.3	1.3	1.4
Emp. Grp/S_{emp}, psf	0.85	1.1	1.1	1.2	1.2	1.2
Ultimate Load Factor, g's	5.7	5.7	5.7	5.7	5.7	5.7
Surface Areas, ft^2						
Wing, S	160	175	175	175	175	174
Horiz. Tail, S_h	28.5	34.6	34.6	34.6	34.1	35.2
Vert. Tail, S_v	14.1	18.4	18.4	18.4	18.4	18.4
Empenn. Area, S_{emp}	42.6	53.0	53.0	53.0	52.5	53.6

*Military observation airplane
**Taildragger

Table A2.2a Group Weight Data for Single Engine Propeller Driven Airplanes

Type Weight Item, lbs.	Cessna 210A	Beech J-35	Saab Safir	Rockwell 112TCA	Cessna 210J
Wing Group	261	379	276	334	335
Empennage Group	71	58	60	98	86
Fuselage Group	316*	200	386	358	408*
Nacelle Group	31	62	in fus.	61	28
Landing Gear Group	207	205	119	161	191
Nose Gear				35	50
Main Gear				126	141
Structure Total	886	904	841	1,082	1,048
Engine	390	432		475	450
Air Induct. System		3			7
Fuel System		30		17	24
Propeller Install.		73		in eng.	64
Engine Install.		45		65	36
Power Plant Total	577	583		557	581
Avionics + Instrum.	16	16		64	18
Surface Controls	44	56	in fus.	44	48
Hydraulic System	4			10	51
Electrical System	60	72		81	57
Air Cond. System Anti-icing System	12	12		in misc.	10
Furnishings	116	174		179	130
Oxygen System	0	0	0	20	0
Ballast	0	0	0	21	0
Auxiliary Gear	0	4	0	2	0
Misc. Equipment	20	0	0	24	0
Paint					21
Fixed Equipm't Total	272	334		445	335
$W_{oil} + W_{tof}$		11		31	24
Fuel (max. payload)		234		230	464**
Payload		845		740	693

*Includes wing-fuselage carry-through spars
**Maximum fuel

Table A2.2b Group Weight Data for Single Engine Propeller Driven Airplanes

Type	Cessna 210A	Beech J-35	Saab Safir	Rockwell 112TCA	Cessna 210J
Flight Design Gross Weight, GW, lbs	2,900	2,900	2,660	2,954	3,400
Structure/GW	0.306	0.312	0.316	0.366	0.308
Power Plant/GW	0.199	0.201		0.189	0.171
Fixed Equipm't/GW	0.094	0.115		0.151	0.099
Empty Weight/GW	0.598	0.628	0.620	0.705	0.578
Wing Group/GW	0.090	0.131	0.104	0.113	0.099
Empenn. Group/GW	0.024	0.020	0.023	0.033	0.025
Fuselage Group/GW	0.109	0.069	0.145	0.121	0.120
Nacelle Group/GW	0.011	0.021		0.021	0.008
Land. Gear Group/GW	0.071	0.071	0.045	0.055	0.056
Take-off Gross Wht, W_{TO}, lbs	2,900	2,900	2,660	2,954	3,400
Empty Weight, W_E, lbs	1,735	1,821	1,650	2,084	1,964
Wing Group/S, psf	1.5	2.1	1.9	2.2	1.9
Emp. Grp/S_{emp}, psf	1.3	1.6	1.4	2.0	1.5
Ultimate Load Factor, g's	5.7				5.7
Surface Areas, ft^2					
Wing, S	176	178	146	152	176
Horiz. Tail, S_h	38.6	*	27.6**	32.0	38.6
Vert. Tail, S_v	17.2	*	14.3**	17.0	17.2
Empenn. Area, S_{emp}	55.8	35.8	41.9	49.0	55.8

*V-tail
**Estimated

Table A3.1a Group Weight Data for Twin Engine Propeller Driven Airplanes

Type	Beech 65 QA*	E-18S	G-50 TB*	95 TA*	Cessna 310C
Number of engines:	2	2	2	2	2
Weight Item, lbs					
Wing Group	670	874	656	458	453
Empennage Group	153	180	156	79	118
Fuselage Group	601	768	495	276	319
Nacelle Group	285	331	261	180	129
Landing Gear Group	444	585**	447	218	263
Nose Gear					
Main Gear					
Structure Total	2,153	2,738	2,015	1,211	1,282
Engines	1,008	1,352	1,008	519	852
Air Induct. System	27	149	27	8	7
Fuel System	137	274	137	83	76
Propeller Install.	258	334	258	162	162
Engine Install.	180	172	172	101	153
Power Plant Total	1,610	2,281	1,610	873	1,250
Avionics + Instrum.	70	100	80	49	46
Surface Controls	132	115	120	73	66
Electrical System	166	295	184	96	121
Electronics	2	63	9	26	0
Air Cond. System Anti-icing System	90	144	81	48	46
Furnishings	438	524	333	194	154
Auxiliary Gear	5	0	7	0	65
Fixed Equipm't Total	903	1,241	814	486	498
$W_{oil} + W_{tfo}$	60	128	60	30	45
Fuel	1,380	1,908	1,380	672	612
Payload	1,287	1,474	1,311	733	1,186

*QA = Queen Air, TB = Twin Bonanza, TA = Travel Air
**Taildragger

Table A3.1b Group Weight Data for Twin Engine Propeller Driven Airplanes

Type	Beech 65 QA*	E-18S**	G-50 TB*	95 TA*	Cessna 310C
Flight Design Gross Weight, GW, lbs	7,368	9,700	7,150	4,000	4,830
Structure/GW	0.292	0.282	0.282	0.303	0.265
Power Plant/GW	0.219	0.235	0.225	0.218	0.259
Fixed Equipm't/GW	0.123	0.128	0.114	0.122	0.103
Empty Weight/GW	0.638	0.651	0.624	0.649	0.628
Wing Group/GW	0.091	0.090	0.092	0.115	0.094
Empenn. Group/GW	0.021	0.019	0.022	0.020	0.024
Fuselage Group/GW	0.082	0.079	0.069	0.069	0.066
Nacelle Group/GW	0.039	0.034	0.037	0.045	0.027
Land. Gear Group/GW	0.060	0.060	0.063	0.055	0.054
Take-off Gross Wht, W_{TO}, lbs	7,368	9,700	7,150	4,000	4,830
Empty Weight, W_E, lbs	4,701	6,318	4,459	2,595	3,032
Wing Group/S, psf	2.4	2.4	2.4	2.4	2.6
Emp. Grp/S_{emp}, psf	1.4	1.7	1.4	1.2	1.5
Ultimate Load Factor, g's	6.6		7.1		5.7
Surface Areas, ft^2					
Wing, S	277	361	277	194	175
Horiz. Tail, S_h	79.3	71.6	79.3	42.4	54.3
Vert. Tail, S_v	30.8	33.6	30.8	23.3	25.9
Empenn. Area, S_{emp}	110	105	110	65.7	80.2

*QA = Queen Air, TB = Twin Bonanza, TA = Travel Air
**Taildragger

Table A3.2a Group Weight Data for Twin Engine Propeller Driven Airplanes

Type	Cessna 404-3	414A	TP-441	Rockwell 690B
Number of engines:	2	2	2	2
Weight Item, lbs	(PP)	(PP)	(PP)	(TBP)
Wing Group	860	638	873	1,001
Empennage Group	181	160	233	207
Fuselage Group	610	678	873	1,377
Nacelle Group	284	200	258	in prop/eng
Land. Gear Group	316	303	346	437
Nose Gear	67	75	69	53
Main Gear	249	228	277	384
Structure Total	2,251	1,979	2,583	3,022
Engines	1,000	862	745	720
Air Induct. System	23	36	0	17
Fuel System	107	96	93	180
Propeller Install.	215	165	302	758
Engine Install.	281	240	130	
Power Plant Total	1,626	1,399	1,270	1,675
Avionics + Instrum.	311	334	250	344
Hydraulic System	52	14	49	99
Surface Controls	113	107	223	81
Electrical System	169	157	403	379
Electronics	1	1	150	0
Oxygen System	0	0	0	23
Air Cond. System	49	130*	182*	205*
Anti-icing System	11	3	78	84
Furnishings	370	342	538	612
Auxiliary Gear	5	3	7	41**
Paint	48	42	48	40
Fixed Equipm't Total	1,129	1,133	1,928	1,908
$W_{oil} + W_{tfo}$	116	113	98	65
Fuel	1,379	961	2,446	1,575
Payload***	1,900	1,200	1,600	1,960

*Includes pressurization system
**This is all ballast in this model
***Includes a crew of two

Table A3.2b Group Weight Data for Twin Engine Propeller Driven Airplanes

Type	Cessna 404-3 (PP)	414A (PP)	TP-441 (TBP)	Rockwell 690B (TBP)
Flight Design Gross Weight, GW, lbs	8,400	6,785	9,925	10,205
Structure/GW	0.268	0.292	0.260	0.296
Power Plant/GW	0.194	0.206	0.128	0.164
Fixed Equipm't/GW	0.134	0.167	0.194	0.187
Empty Weight/GW	0.596	0.665	0.582	0.647
Wing Group/GW	0.102	0.094	0.088	0.098
Empenn. Group/GW	0.022	0.024	0.023	0.020
Fuselage Group/GW	0.073	0.100	0.088	0.135
Nacelle Group/GW	0.034	0.029	0.026	
Land. Gear Group/GW	0.038	0.045	0.035	0.043
Take-off Gross Wht, W_{TO}, lbs	8,400	6,785	9,925	10,205
Empty Weight, W_E, lbs	5,006	4,511	5,781	6,605
Wing Group/S, psf	3.6	2.8	3.4	3.8
Emp. Grp/S_{emp}, psf	1.7	1.6	2.2	2.0
Ultimate Load Factor, g's	3.75*	3.75*	3.75*	3.75*
Surface Areas, ft^2				
Wing, S	242	226	254	266
Horiz. Tail, S_h	63.4	60.7	63.4	58.4
Vert. Tail, S_v	43.5	41.2	43.5	44.8
Empenn. Area, S_{emp}	107	102	107	103

*Assumed

Table A4.1a Group Weight Data for Agricultural Airplanes

Type	PZL M18 Dromader		PZL M21 Dromader Mini		PZL M15 Belphegor	
Number of Engines	1 Piston		1 Piston		1 Turbofan	
Weight Item, kg (lbs)	kg	lbs	kg	lbs	kg	lbs
Wing Group	731	1612	660	1455	893	1969
Empennage Group	86	190	82	181	146	322
Fuselage Group	280	617	245	540	402	866
Nacelle Group	66	146	48	106	402	866
Landing Gear Group	188	415	146	322	229	505
Nose Gear	-	-	-	-	64	141
Tail Gear	24	53	24	53	-	-
Main Gear	164	362	122	269	165	364
Structural Total	1351	2980	1181	2604	1670	3682
Engines	567	1250	464	1023	355	783
Propellers	188	414	105	231	-	-
Air Induction System	23	51	12	26	39	86
Exhaust System	19	42	17	37	6	13
Fuel System	21	46	18	40	110	243
Oil System	46	101	40	88	-	-
Propulsion system (engine, propell. controls, engine starting system, accessory cooling, miscellaneous)	29	64	12	26	22	48
Power Plant Total	893	1968	668	1471	532	1173
Fixed Agro. Equim't:	157	346	105	231	420	926
Hopper	140	309	90	198	-	-
Fixed Argo System. (controls, loading inst., etc.)	17	37	15	33	-	-
Surface Controls	40	88	36	79	78	172
Avionics & Instruments	16	35	16	35	15	33
Hydraulic System	29	64	29	64	77	170
Pneumatic System	-	-	-	-	-	-
Electrical System	88	194	85	187	173	381
Electronics	5	11	5	11	13	29
Venting & Air Cond. Sys.	6	13	6	13	21	46
Furnishings	25	55	25	55	29	64
Miscellaneous	25	55	20	44	-	-
Fixed Equipm't Total	391	861	327	719	826	1821
Spray Equipment	134	295	125	276	103	227
Spreading Equipment	129	284	92	203	64	141
Atomizing Equipment	181	399	163	359	107	236
Fire Fighting Equipment	60	132	60	132	-	-
Agro Operational Equipm't (one among a/m)	60 132	181 400	60 132	163 360	64 141	107 236
W_{oil}	60	132	60	132	23	51
W_{tfo}	7	15	6	13	6	13
Maximum Fuel Capacity	1300	2866	900	1984	2100	4630
Maximum Payload	2200	4850	1100	2425	2200	4850

Table A4.1b Group Weight Data for Agricultural Airplanes

Type	At the time of printing no data were available
Flight Design Gross Weight, GW, lbs	
Structure/GW	
Power Plant/GW	
Fixed Equipm't/GW	
Empty Weight/GW	
Wing Group/GW	
Empenn. Group/GW	
Fuselage Group/GW	
Nacelle Group/GW	
Land. Gear Group/GW	
Take-off Gross Wht, W_{TO}, lbs	
Empty Weight, W_E, lbs	
Wing Group/S, psf	
Emp. Grp/S_{emp}, psf	
Ultimate Load Factor, g's	
Surface Areas, ft^2	
Wing, S	
Horiz. Tail, S_h	
Vert. Tail, S_v	
Empenn. Area, S_{emp}	

Table A5.1a Group Weight Data for Business Jets

Type	MS-760 Paris	Lockheed Jetstar	Gates-Learjet 25D	Gates-Learjet 28
Number of engines:	2	2	2	2
Weight Item, lbs				
Wing Group	897	2,827	1,467	1,939
Empennage Group	176	879	361	361
Fuselage Group	912	3,491	1,575	1,624
Nacelle Group	49*	792	241	214
Landing Gear Group	307	1,061	584	584
Nose Gear			102	102
Main Gear			482	482
Structure Total	2,341	9,050	4,228	4,722
Engines	609	1,750	792	792
Air Induct. System	31	135	0	0
Fuel System	240	360	179	237
Propulsion System.	136	230	255	255
Power Plant Total	1,016	2,475	1,226	1,284
Avionics + Instrum.	70	153	383	383
Surface Controls	188	768	291	275
Hydraulic System Pneumatic System		262	119	114
Electrical System	284	973	620	603
Electronics	158	868	0	0
Oxygen System			28	26
Air Cond. System**	48	510	293	285
Anti-icing System			82	162
Furnishings	169	1,521	720	768
Auxiliary Gear	0	10	0	0
Miscellaneous	0	0	-40	-11
Fixed Equipm't Total	917	5,065	2,496	2,605
$W_{oil} + W_{tfo}$	28	204	177	
Max. Fuel Capacity	2,460	11,229	6,098	4,684
Max. Payload	884	2,100	2,980	1,962

*Engines buried inside the fuselage
**Includes pressurization system

Table A5.1b Group Weight Data for Business Jets

Type	MS-760 Paris	Lockheed Jetstar	Gates-Learjet 25D	Gates-Learjet 28
Flight Design Gross Weight, GW, lbs	7,650	30,680	15,000	15,000
Structure/GW	0.306	0.295	0.282	0.315
Power Plant/GW	0.133	0.081	0.082	0.086
Fixed Equipm't/GW	0.120	0.165	0.166	0.174
Empty Weight/GW*	0.563	0.541	0.530	0.574
Wing Group/GW	0.117	0.092	0.098	0.129
Empenn. Group/GW	0.023	0.029	0.024	0.024
Fuselage Group/GW	0.119	0.114	0.105	0.108
Nacelle Group/GW	0.006*	0.026	0.016	0.014
Land. Gear Group/GW	0.040	0.035	0.039	0.039
Take-off Gross Wht, W_{TO}, lbs	7,650	30,680	15,000	15,000
Empty Weight, W_E, lbs	4,306	16,590	7,950	8,611
Wing Group/S, psf	4.6	5.4	6.3	7.3
Emp. Grp/S_{emp}, psf	3.5	3.4	3.9	3.9
Ultimate Load Factor, g's	3.75**	5.25	3.75**	3.75**
Surface Areas, ft^2				
Wing, S	194	521	232	265
Horiz. Tail, S_h	31.8	149	54.0	54.0
Vert. Tail, S_v	18.4	110	37.4	37.4
Empenn. Area, S_{emp}	50.2	259	91.4	91.4

*Engines buried inside the fuselage
**Assumed

Table A5.2a Group Weight Data for Business Jets

Type	Cessna Citation II	Rockwell JC-1121	Hawker-Siddeley 125	Gulfstr. American GII
Number of engines:	2	2	2	2
Weight Item, lbs				
Wing Group	1,288	1,322	1,968	6,372
Empenn. Group	295	425	608	1,965
Fuselage Group	1,069	1,622	1,628	5,944
Nacelle Group	220	350	in fusel.	1,239
Land. Gear Group	465	443	659	2,011
Nose Gear	87			321
Main Gear	378			1,690
Structure Total	3,337	4,162	4,863	17,531
Engine(s)	1,100			6,570
Air Induct. System	26			
Exhaust System	15			
Fuel System	189			316
Propulsion System	105			
Power Plant Total	1,435			6,886
Avionics + Instrum.	87			1,715
Surface Controls	203	223	217	1,021
Hydraulic System	96			959
Electrical System	340			1,682
Electronics	313			
Oxygen System				140
Air Cond. System*	264			927
Anti-icing System	98			
Furnishings	800			4,501
Auxiliary Gear	3			
Auxiliary power unit				258
Paint	47			
Fixed Equipm't Total	2,251			11,203
$W_{oil} + W_{tfo}$	143			
Max. Fuel Capacity	5,009	8,964	9,193	23,300
Max. Payload			1,905	5,380

*Includes pressurization system

Table A5.2b Group Weight Data for Business Jets

Type	Cessna Citation II	Rockwell JC-1121	Hawker Siddeley 125	Gulfstr. American GII
Flight Design Gross Weight, GW, lbs	13,500	20,500	23,300	64,800
Structure/GW	0.247	0.203	0.209	0.271
Power Plant/GW	0.106			0.106
Fixed Equipm't/GW	0.167			0.173
Empty Weight/GW	0.520	0.540	0.526	0.550*
Wing Group/GW	0.095	0.064	0.084	0.098
Empenn. Group/GW	0.022	0.021	0.026	0.030
Fuselage Group/GW	0.079	0.079	0.070	0.092
Nacelle Group/GW	0.016	0.017	in fusel.	0.019
Land. Gear Group/GW	0.034	0.022	0.028	0.031
Take-off Gross Wht, W_{TO}, lbs	13,500	20,500	23,300	64,800
Empty Weight, W_E, lbs	7,023	11,070	12,260	35,620
Wing Group/S, psf	4.6	4.4		8.0
Emp. Grp/S_{emp}, psf	2.4	3.3		5.8
Ultimate Load Factor, g's	3.75**			
Surface Areas, ft^2				
Wing, S	279	303	353	794
Horiz. Tail, S_h	70.6	70	100	182
Vert. Tail, S_v	50.9	59.3	51.6	155
Empenn. Area, S_{emp}	122	129	152	337

*Typical. Individual airplanes will vary.   **Assumed

Table A6.1a Group Weight Data for Regional Turbopropeller Driven Airplanes

Type	Grumman G-I	Fokker F-27-100	Nord 262	Embraer 110-P2
Number of engines:	2	2	2	2
Weight Item, lbs				
Wing Group	3,735	4,408	2,698	1,502
Empennage Group	874	977	805	454
Fuselage Group	3,718	4,122	3,675	1,354
Nacelle Group	1,136	628	236	198
Land. Gear Group	1,207	1,940	1,085	538
Nose Gear	219			
Main Gear	988			
Structure Total	10,670	12,075	8,499	4,046
Engines	2,688	2,427		622
Air Induct. System				
Fuel System	133	390		86
Propeller Install.	1,002	918		1,140
Propulsion System	698	612		in prop.
Power Plant Total	4,521	4,347		1,848
Avionics + Instrum.	97	81	133	364
Surface Controls	461	613	408	342
Hydraulic System Pneumatic System	235	242	(incl.in electr.)	176
Electrical System	966	835	765	452
Electronics	99	386	238	in avion.
APU	355	0	0	0
Air Cond. System Anti-icing System	755*	1,225*	527*	192 73
Furnishings	415	2,291	1,324	882
Auxiliary Gear	6		33	
Fixed Equipm't Total	3,389	5,673	3,428	2,481
$W_{oil} + W_{tfo}$	329			
Max. Fuel Capacity	10,447	9,198	3,559	3,062
Maximum Payload	4,270	12,500	6,175	3,706

*Includes pressurization system

Table A6.1b Group Weight Data for Regional Turbopropeller Driven Airplanes

Type	Grumman G-I	Fokker F-27-100	Nord 262	Embraer 110-P2
Flight Design Gross Weight, GW, lbs	35,100	37,500	22,930	12,500
Structure/GW	0.304	0.322	0.371	0.324
Power Plant/GW	0.129	0.116		0.148
Fixed Equipm't/GW	0.097	0.151	0.149	0.198
Empty Weight/GW	0.624	0.615	0.663	0.670
Wing Group/GW	0.106	0.118	0.118	0.120
Empenn. Group/GW	0.025	0.026	0.035	0.036
Fuselage Group/GW	0.106	0.110	0.160	0.108
Nacelle Group/GW	0.032	0.017	0.010	0.016
Land. Gear Group/GW	0.034	0.052	0.047	0.043
Take-off Gross Wht, W_{TO}, lbs	35,100	37,500	22,930	12,500
Empty Weight, W_E, lbs	21,900	23,054	15,200	8,375
Wing Group/S, psf	6.1	5.8	4.6	4.8
Emp. Grp/S_{emp}, psf	3.6	3.0	2.9	2.8
Ultimate Load Factor, g's	3.75*	3.75*	3.75*	3.75*
Surface Areas, ft^2				
Wing, S	610	754	592	313
Horiz. Tail, S_h	127	172	169	105
Vert. Tail, S_v	117	153	109	59
Empenn. Area, S_{emp}	244	325	278	164

*Assumed

Table A6.2a Group Weight Data for Regional Turbopropeller Driven Airplanes

Type	Fokker F-27-200	F-27-500	Short* Skyvan
Number of engines:	2	2	2
Weight Item, lbs			
Wing Group	4,505	4,510	1,220
Empennage Group	1,053	1,060	374
Fuselage Group	4,303	5,142	2,154
Nacelle Group	667	668	254
Land. Gear Group	1,825	1,865	466
Nose Gear			
Main Gear			
Structure Total	12,353	13,245	4,468
Engines			714
Air Induct. System			
Fuel System			373
Propeller Install.			368
Propulsion System			87
Power Plant Total			1,542
Avionics + Instrum.		126	74
Surface Controls	620	626	265
Hydraulic System		256	64
Pneumatic System			
Electrical System		840	320
Electronics		329	12
APU		0	
Air Cond. System		1,257**	85
Anti-icing System			
Furnishings		3,035	135***
Auxiliary Gear		0	43
Paint			75
Fixed Equipm't Total		6,469	1,073
$W_{oil} + W_{tfo}$			44
Max. Fuel Capacity	9,146	9,146	4,924
Payload (Max.)	12,615	12,383	

*Strutbraced wing **Includes pressurization
***Cockpit furnishings only

Table A6.2b Group Weight Data for Regional Turbopropeller Driven Airplanes

Type	Fokker F-27-200	F-27-500	Short* Skyvan
Flight Design Gross Weight, GW, lbs	43,500	45,000	12,500
Structure/GW	0.284	0.294	0.357
Power Plant/GW			0.123
Fixed Equipm't/GW		0.144	0.086
Empty Weight/GW	0.537	0.548	0.570
Wing Group/GW	0.104	0.100	0.098
Empenn. Group/GW	0.024	0.024	0.030
Fuselage Group/GW	0.099	0.114	0.172
Nacelle Group/GW	0.015	0.015	0.020
Land. Gear Group/GW	0.042	0.041	0.037
Take-off Gross Wht, W_{TO}, lbs	43,500	45,000	12,500
Empty Weight, W_E, lbs	23,350	24,650	7,125
Wing Group/S, psf	6.0	6.0	3.3
Emp. Grp/S_{emp}, psf	3.2	3.3	2.2
Ultimate Load Factor, g's	3.75**	3.75**	3.75**
Surface Areas, ft^2			
Wing, S	754	754	373
Horiz. Tail, S_h	172	172	85
Vert. Tail, S_v	153	153	83
Empenn. Area, S_{emp}	325	325	168

*Strutbraced wing
**Assumed

Table A6.3a Group Weight Data for Regional Turbopropeller Driven Airplanes

Type	De Havilland Canada	
	DHC7-102	DHC6-300
Number of engines:	2	2
Weight Item, lbs		
Wing Group	4,888	1,263*
Empennage Group	1,318	303
Fuselage Group	4,680	1,705
Nacelle Group	1,841	221
Land. Gear Group	1,732	613
Nose Gear		
Main Gear		
Structure Total	14,459	4,105
Engines		
Air Induct. System		
Fuel System		
Propeller Install.		
Propulsion System		
Power Plant Total	4,701	1,248
Avionics + Instrum.	850	371
Surface Controls	710	145
Hydraulic System Pneumatic System	493	43
Electrical System	1,651	356
Electronics		
Air Cond. System Pressurization System	550	103**
Anti-icing System	176	
Furnishings	2,862	732
Paint	150	64
Fixed Equipm't Total	7,442	1,814
$W_{oil} + W_{tfo}$	150	35
Full oil	130	54
Max. Fuel Capacity	6,968	1,114
Water and supplies	130	
Payload (Max.)	9,500	3,610

*Strutbraced wing **Heating system only

Table A6.3b Group Weight Data for Regional Turbopropeller Driven Airplanes

Type	De Havilland Canada DHC7-102	DHC6-300
Flight Design Gross Weight, GW, lbs	44,000	12,500
Structure/GW	0.329	0.328
Power Plant/GW	0.107	0.100
Fixed Equipm't/GW	0.169	0.145
Empty Weight/GW	0.605	0.573
Wing Group/GW	0.111	0.101*
Empenn. Group/GW	0.030	0.024
Fuselage Group/GW	0.106	0.136
Nacelle Group/GW	0.042	0.018
Land. Gear Group/GW	0.039	0.049**
Take-off Gross Wht, W_{TO}, lbs	44,000	12,500
Empty Weight, W_E, lbs	26,602	7,167
Wing Group/S, psf	5.7	3.0
Emp. Grp/S_{emp}, psf	3.4	2.1
Ultimate Load Factor, g's	3.75***	3.75***
Surface Areas, ft^2		
Wing, S	860	420
Horiz. Tail, S_h	217	100
Vert. Tail, S_v	170	48
Empenn. Area, S_{emp}	387	148

*Strutbraced wing **Fixed gear
***Assumed

Table A6.4a Group Weight Data for Regional Piston/ Propeller Driven Airplanes

Type	SAAB Scandia	Handley Page Herald	Scottish* Aviation Twin Pion.	Convair 240
Number of engines:	2	4	2	2
Weight Item, lbs				
Wing Group	4,195	4,365	2,121	3,943
Empennage Group	584	987	576	922
Fuselage Group	2,773	2,986	1,381	4,227
Nacelle Group	1,479	830	230	1,215
Land. Gear Group	1,841	1,625	703	1,530
Nose Gear				
Main Gear				
Structure Total	10,872	10,793	5,011	11,837
Engines				
Air Induct. System				
Fuel System				
Propeller Install.				
Propulsion System				
Power Plant Total				7,299
Avionics + Instrum.				
Surface Controls	369	364	300	
Hydraulic System				
Pneumatic System				
Electrical System				
Electronics				
APU				
Oxygen System				
Air Cond. System				
Anti-icing System				
Furnishings				
Auxiliary Gear				
Fixed Equipm't Total				4,444
$W_{oil} + W_{tfo}$	not----------	----------	----------	----------known
Max. Fuel Capacity	11,080		1,740	6,700
Payload (Max.)			2,950	16,000

*Strutbraced wing

Table A6.4b Group Weight Data for Regional Piston/ Propeller Driven Airplanes

Type	SAAB Scandia	Handley Page Herald	Scottish Aviation Twin Pion.	Convair 240
Flight Design Gross Weight, GW, lbs	30,860	37,500	14,600	43,500
Structure/GW	0.352	0.288	0.343	0.272
Power Plant/GW				0.168
Fixed Equipm't/GW				0.102
Empty Weight/GW	0.641	0.673	0.683	0.542
Wing Group/GW	0.136	0.116	0.145	0.091
Empenn. Group/GW	0.019	0.026	0.039	0.021
Fuselage Group/GW	0.090	0.080	0.095	0.097
Nacelle Group/GW	0.048	0.022	0.016	0.028
Land. Gear Group/GW	0.060	0.043	0.048	0.035
Take-off Gross Wht, W_{TO}, lbs	30,860	37,500	14,600	43,500
Empty Weight, W_E, lbs	19,780	25,240	9,969	23,580
Wing Group/S, psf	4.5	4.9	3.2	4.8
Emp. Grp/S_{emp}, psf	2.0	2.2	1.7	
Ultimate Load Factor, g's	3.75*	3.75*	3.75*	3.75*
Surface Areas, ft^2				
Wing, S	922	886	670	817
Horiz. Tail, S_h	215	252	167	
Vert. Tail, S_v	82	193	167	
Empenn. Area, S_{emp}	297	445	334	

*Assumed

Table A7.1a Group Weight Data for Jet Transports

Type	McDonnell Douglas DC-9-30	MD-80	DC-10-10	DC-10-30
Number of engines:	2	2	3	3
Weight Item, lbs				
Wing Group	11,400	15,560	48,990	58,859
Empennage Group	2,780	3,320	13,660	14,676
Fuselage Group	11,160	16,150	44,790	47,270
Nacelle Group	1,430	2,120	8,490	9,127
Land. Gear Group	4,170	5,340	19,820	25,761
Nose Gear	470	550	1,520	1,832
Main Gear	3,700	4,790	18,300	23,929*
Structure Total	30,940	42,490	135,750	155,693
Engine(s)	6,410	8,820	23,688	26,163
Exhaust and Thrust Reverser System	1,240	1,540	7,232	6,916
Air Induct. System	0	0	0	0
Fuel System	600	640	2,040	4,308
Propulsion Install.	0	0	0	0
Power Plant Total	8,250	11,000	32,960	37,387
Avionics + Instrum.	1,450	2,130	3,410	4,274
Surface Controls	1,620	2,540	5,880	6,010
Hydraulic System	480	540	2,330	2,587
Pneumatic System	280	290	1,790	1,920
Electrical System	1,330	1,720	5,370	5,912
Electronics	Included in Avionics and Instrum.			
APU	820	840	1,590	1,643
Oxygen System	150	220	210	256
Air Cond. System**	1,120	1,580	2,390	2,723
Anti-icing System	480	550	420	471
Furnishings	8,450	11,400	35,810	34,124
Operating Items	2,700	3,650	13,340	16,274
Fixed Equipm't Total	18,880	25,460	72,540	76,194
W_{tfo}	Not known			
Max. Fuel Capacity	28,746	39,362	146,683	247,034
Max. Payload	28,930	43,050	93,750	98,726

*Includes 3,590 lbs for centerline gear
**Includes pressurization system

Table A7.1b Group Weight Data for Jet Transports

Type	Mc Donnell Douglas DC-9-30	MD-80	DC-10-10	DC-10-30
Flight Design Gross Weight, GW, lbs	108,000	140,000	430,000	555,000
Structure/GW	0.286	0.304	0.316	0.281
Power Plant/GW	0.076	0.079	0.077	0.067
Fixed Equipm't/GW	0.175	0.182	0.169	0.137
Empty Weight/GW	0.538	0.564	0.561	0.485
Wing Group/GW	0.106	0.111	0.114	0.106
Empenn. Group/GW	0.026	0.024	0.032	0.026
Fuselage Group/GW	0.103	0.115	0.104	0.085
Nacelle Group/GW	0.013	0.015	0.020	0.016
Land. Gear Group/GW	0.039	0.038	0.046	0.046
Take-off Gross Wht, W_{TO}, lbs	108,000	140,000	430,000	555,000
Empty Weight, W_E, lbs	58,070	78,950	241,250	269,274
Wing Group/S, psf	11.4	12.3	12.7	14.9
Emp. Grp/S_{emp}, psf	6.4	5.7	7.0	7.6
Ultimate Load Factor, g's	3.75*	3.75*	3.75*	3.75*
Surface Areas, ft^2				
Wing, S	1,001	1,270	3,861	3,958
Horiz. Tail, S_h	276	314	1,338	1,338
Vert. Tail, S_v	161	168	605	605
Empenn. Area, S_{emp}	437	582	1,943	1,943

*Assumed

Table A7.2a Group Weight Data for Jet Transports

Type	Boeing 737-200	727-100	747-100	Airbus A-300 B2
Number of engines:	2	3	4	2
Weight Item, lbs				
Wing Group	10,613	17,764	86,402	44,131
Empennage Group	2,718	4,133	11,850	5,941
Fuselage Group	12,108	17,681	71,845	35,820
Nacelle Group	1,392	3,870	10,031	7,039
Land. Gear Group	4,354	7,211	31,427	13,611
Nose Gear				
Main Gear				
Structure Total	31,185	50,659	211,555	106,542
Engines	6,217	9,325	34,120	16,825
Exhaust and Thrust-Reverser System	1,007	1,744	6,452	4,001
Air Induct. System	0	0	0	0
Fuel System	575	1,143	2,322	1,257
Propulsion Install.	378	250	802	814
Power Plant Total	8,177	12,462	43,696	22,897
Avionics + Instrum.	625	756	1,909	377
Surface Controls	2,348	2,996	6,982	5,808
Hydraulic System Pneumatic System	873	1,418	4,471	3,701
Electrical System	1,066	2,142	3,348	4,923
Electronics	956	1,591	4,429	1,726
APU	836	60	1,130	983
Air Cond. System* Anti-icing System	1,416	1,976	3,969	3,642
Furnishings	6,643	10,257	37,245	13,161
Miscellanous	124	85	-421	732
Fixed Equipm't Total	14,887	21,281	63,062	35,053
$W_{oil} + W_{tfo}$	Not	-------------	-------------	known
Max. Fuel Capacity	34,718	48,353	331,675	76,512
Max. Payload	34,790	29,700	140,000	69,865

*Includes pressurization system

Table A7.2b Group Weight Data for Jet Transports

Type	Boeing 737-200	727-100	747-100	Airbus A300-B2
Flight Design Gross Weight, GW, lbs	115,500	160,000	710,000	302,000
Structure/GW	0.270	0.317	0.298	0.353
Power Plant/GW	0.071	0.078	0.062	0.076
Fixed Equipm't/GW	0.129	0.133	0.089	0.116
Empty Weight/GW	0.521	0.552	0.498	0.559
Wing Group/GW	0.092	0.111	0.122	0.146
Empenn. Group/GW	0.024	0.026	0.017	0.020
Fuselage Group/GW	0.105	0.111	0.101	0.119
Nacelle Group/GW	0.012	0.024	0.014	0.023
Land. Gear Group/GW	0.038	0.045	0.044	0.045
Take-off Gross Wht, W_{TO}, lbs	115,500	160,000	710,000	302,000
Empty Weight, W_E, lbs	60,210	88,300	353,398	168,805
Wing Group/S, psf	10.8	10.4	15.7	15.8
Emp. Grp/S_{emp}, psf	4.9	5.6	5.2	4.8
Ultimate Load Factor, g's	3.75*	3.75*	3.75*	3.75*
Surface Areas, ft^2				
Wing, S	980	1,700	5,500	2,799
Horiz. Tail, S_h	321	376	1,470	748
Vert. Tail, S_v	233	356	830	487
Empenn. Area, S_{emp}	554	732	2,300	1,235

*Assumed

Table A7.3a Group Weight Data for Jet Transports

Type	Boeing 707-121	707-320	707-320C	720-022
Number of engines:	4	4	4	4
Weight Item, lbs				
Wing Group	24,024	29,762	32,255	22,850
Empennage Group	5,151	5,511	6,165	5,230
Fuselage Group	20,061	21,650	26,937	19,035
Nacelle Group	4,639	4,497	4,183	4,510
Land. Gear Group	9,763	12,700	12,737	8,110
Nose Gear				
Main Gear				
Structure Total	63,638	74,120	82,277	59,735
Engines	16,458	20,200	17,368	13,770
Exhaust and Thrust-Reverser System			3,492	
Air Induct. System	0	0	0	0
Fuel System	1,808		2,418	1,240
Propulsion Install.	1,738		798	885
Power Plant Total	20,004		24,076	15,895
Avionics + Instrum.	505		515	555
Surface Controls	2,159	2,400	3,052	2,450
Hydraulic System Pneumatic System	484		1,086	505
Electrical System	3,772		4,179	4,070
Electronics	1,708		2,338	1,200
APU			151	
Air Cond. System* Anti-icing System	3,110		3,608	2,890
Furnishings	13,651		9,527	13,055
Auxiliary Gear	0	0	0	0
Miscellanous	0	0	-389	0
Fixed Equipm't Total	25,389		24,456	24,725
W_{tfo}	704			
Max. Fuel Capacity	90,842	160,783	160,783	99,954
Max. Payload	42,600	55,000	84,000	28,200

*Includes pressurization system

Table A7.3b Group Weight Data for Jet Transports

Type	Boeing 707-121	707-320	707-320C	720-022
Flight Design Gross Weight, GW, lbs	246,000	311,000	330,000	203,000
Structure/GW	0.259	0.238	0.249	0.294
Power Plant/GW	0.081		0.073	0.078
Fixed Equipm't/GW	0.103		0.074	0.122
Empty Weight/GW	0.444	0.434	0.396	0.494
Wing Group/GW	0.098	0.096	0.098	0.113
Empenn. Group/GW	0.021	0.018	0.019	0.026
Fuselage Group/GW	0.082	0.070	0.082	0.094
Nacelle Group/GW	0.019	0.014	0.013	0.022
Land. Gear Group/GW	0.040	0.041	0.039	0.040
Take-off Gross Wht, W_{TO}, lbs	246,000	311,000	330,000	203,000
Empty Weight, W_E, lbs	109,111	135,000	130,809	100,355
Wing Group/S, psf	9.9	10.3	10.6	9.4
Emp. Grp/S_{emp}, psf	6.2	5.8	6.5	6.3
Ultimate Load Factor, g's	3.75	3.75	3.75	3.75
Surface Areas, ft^2				
Wing, S	2,433	2,892	3,050	2,433
Horiz. Tail, S_h	500	625	625	500
Vert. Tail, S_v	328	328	328	328
Empenn. Area, S_{emp}	828	953	953	828

Table A7.4a Group Weight Data for Jet Transports

Type	Boeing 707-321	McDonnell Douglas DC-8	DC-9-10	Hawker-Siddeley 121-IC
Number of engines:	4	4	2	3
Weight Item, lbs				
Wing Group	28,647	27,556	9,470	12,600
Empennage Group	6,004	4,840	2,630	3,225
Fuselage Group	22,129	19,910	11,206	12,469
Nacelle Group	5,119	3,534	1,417	in fusel.
Land. Gear Group	11,122	10,910	3,660	4,413
Nose Gear				
Main Gear				
Structure Total	73,021	66,750	28,383	32,707
Engine(s)	19,192		6,160	
Exhaust and Thrust Reverser System			658	
Air Induct. System				
Fuel System	1,956		510	
Propulsive Install.	1,113		409	
Power Plant Total	22,261	27,677	7,737	
Avionics + Instrum.	561		719	
Surface Controls	2,408		1,264	1,792
Hydraulic System / Pneumatic System	498		714	
Electrical System	3,959		1,663	
Electronics	1,716		914	
APU			818	
Air Cond. System* / Anti-icing System	3,290		1,476	
Furnishings	14,854		7,408	
Auxiliary Gear	0		24	
Fixed Equipm't Total	27,286	25,650	15,000	
W_{tfo}	1,089			
Max. Fuel Capacity		30,256	18,778	31,060
Max. Payload			18,050	22,000

*Includes pressurization system

Table A7.4b Group Weight Data for Jet Transports

Type	Boeing 707-321	McDonnell Douglas DC-8	McDonnell Douglas DC-9-10	Hawker Siddeley 121-IC
Flight Design Gross Weight, GW, lbs	302,000	215,000	91,500	115,000
Structure/GW	0.242	0.310	0.310	0.284
Power Plant/GW	0.074	0.129	0.085	
Fixed Equipm't/GW	0.090	0.119	0.164	
Empty Weight/GW	0.406	0.562	0.495	0.587
Wing Group/GW	0.095	0.128	0.103	0.110
Empenn. Group/GW	0.020	0.023	0.029	0.028
Fuselage Group/GW	0.073	0.093	0.122	0.108
Nacelle Group/GW	0.017	0.016	0.015	in fusel.
Land. Gear Group/GW	0.037	0.051	0.040	0.038
Take-off Gross Wht, W_{TO}, lbs	301,000	215,000	91,500	115,000
Empty Weight, W_E, lbs	122,509	120,877	45,300	67,500
Wing Group/S, psf	9.9	9.9	10.1	9.3
Emp. Grp/S_{emp}, psf	6.4	5.6	5.5	5.7
Ultimate Load Factor, g's	3.75	3.75*	3.75*	3.75*
Surface Areas, ft^2				
Wing, S	2,892	2,773	934	1,358
Horiz. Tail, S_h	625	607**	275	310
Vert. Tail, S_v	312	263**	200**	259
Empenn. Area, S_{emp}	937	870	475	569

*Assumed

**Estimated from threeview

Table A7.5a Group Weight Data for Jet Transports

Type	VFW-Fokker 614	Fokker F28-1000	BAC 1-11/300	Sud/Aerospatiale Caravelle
Number of engines:	2	2	2	2
Weight Item, lbs				
Wing Group	5,767	7,330	9,643	14,735
Empennage Group	1,121	1,632	2,369	1,957
Fuselage Group	5,233	7,043	9,713	11,570
Nacelle Group	971	834	in fusel.	1,581
Land. Gear Group	1,620	2,759	2,856	5,110
Nose Gear				
Main Gear				
Structure Total	14,712	19,598	24,581	34,953
Engines	3,413	4,495		7,055
Exhaust and Thrust-Reverser System	119	127		975
Air Induct. System				
Fuel System	162	545		518
Propulsive Install	690	215		179
Power Plant Total	4,384	5,382		8,727
Avionics + Instrum.	215	302	182	236
Surface Controls	745	1,387	1,481	2,063
Hydraulic System Pneumatic System	403	364	997	1,376
Electrical System	1,054	1,023	2,317	2,846
Electronics	436	869	1,005	1,187
APU	305	346	457	0
Air Cond. System* Anti-icing System	719	1,074	1,579	1,752
Furnishings	2,655	4,030	4,933	6,481
Auxiliary Gear	49	0	0	0
Operating Items				
Fixed Equipm't Total	6,581	9,395	12,951	15,941
W_{tfo}	not			known
Max. Fuel Capacity	10,142	17,331	24,954	33,808
Max. Payload	8,201	14,380	22,278	29,100

*Includes pressurization system

Table A7.5b Group Weight Data for Jet Transports

Type	VFW Fokker 614	Fokker F28-1000	BAC 1-11/300	Sud-Aero spatiale Caravelle
Flight Design Gross Weight, GW, lbs	40,981	65,000	87,000	110,230
Structure/GW	0.359	0.302	0.283	0.317
Power Plant/GW	0.107	0.083		0.079
Fixed Equipm't/GW	0.161	0.145	0.149	0.145
Empty Weight/GW*	0.586	0.480	0.560	0.590
Wing Group/GW	0.141	0.113	0.111	0.134
Empenn. Group/GW	0.027	0.025	0.027	0.018
Fuselage Group/GW	0.128	0.108	0.112	0.105
Nacelle Group/GW	0.024	0.013	in fusel.	0.014
Land. Gear Group/GW	0.040	0.042	0.033	0.046
Take-off Gross Wht, W_{TO}, lbs	40,981	65,000	87,000	110,230
Empty Weight, W_E, lbs	24,000	31,219	48,722	65,050
Wing Group/S, psf	8.4	8.9	9.6	9.3
Emp. Grp/S_{emp}, psf	3.8	4.8	6.3	4.2
Ultimate Load Factor, g's	3.75*	3.75*	3.75*	3.75*
Surface Areas, ft^2				
Wing, S	689	822	1,003	1,579
Horiz. Tail, S_h	193	210	257	301
Vert. Tail, S_v	102	132	117	167
Empenn. Area, S_{emp}	295	342	374	468

*Assumed

Table A7.6a Group Weight Data for Turboprop. Transports

Type	Bristol Britannia 300	Canadair CL-44C	Vickers Viscount 810	Lockheed Electra
Number of engines:	4	4	4	4
Weight Item, lbs				
Wing Group	13,433	15,710	6,250	7,670
Empennage Group	3,202	3,749	1,245	1,924
Fuselage Group	11,100	20,524	6,900	9,954
Nacelle Group	4,930	6,834	1,810	4,417
Land. Gear Group	5,785	7,083	2,469	3,817
Nose Gear				
Main Gear				
Structure Total	38,450	53,900	18,674	27,782
Engines	11,192	12,800		
Air Induct. System				
Fuel System	1,329	1,755		
Propeller Inst.	3,557	5,006		
Propulsion System	3,820	3,134		
Power Plant Total	19,898	22,695		13,733
Avionics + Instrum.	505	858	213	
Surface Controls	1,221	2,146	824	
Hydraulic System Pneumatic System	650	630	457	
Electrical System	1,800	3,040	2,826	
Electronics	1,040	1,229	617	
APU	0	0	0	
Air Cond. System* Anti-icing System	3,000	2,536	2,092	
Furnishings	6,866	12,349	3,476	
Auxiliary Gear	0	0	0	
Fixed Equipm't Total	15,082	22,788	10,505	14,469
$W_{oil} + W_{tfo}$				
Max. Fuel Capacity	69,395	82,170	13,897	37,205
Payload (Max.)	30,000	37,630	15,054	18,907

*Includes pressurization system

Table A7.6b Group Weight Data for Turboprop. Transports

Type	Bristol Britannia 300	Canadair CL-44C	Vickers Viscount 810	Lockheed Electra
Flight Design Gross Weight, GW, lbs	155,000	205,000	72,500	116,000
Structure/GW	0.248	0.263	0.258	0.240
Power Plant/GW	0.128	0.111		0.118
Fixed Equipm't/GW	0.097	0.111	0.145	0.125
Empty Weight/GW	0.587	0.516	0.569	0.491
Wing Group/GW	0.087	0.077	0.086	0.066
Empenn. Group/GW	0.021	0.018	0.017	0.017
Fuselage Group/GW	0.072	0.100	0.095	0.086
Nacelle Group/GW	0.032	0.033	0.025	0.038
Land. Gear Group/GW	0.037	0.035	0.034	0.033
Take-off Gross Wht, W_{TO}, lbs	155,000	205,000	72,500	116,000
Empty Weight, W_E, lbs	91,000	105,785	41,276	57,000
Wing Group/S, psf	6.5	7.6	6.5	5.9
Emp. Grp/S_{emp}, psf	3.4	4.0	2.9	3.1
Ultimate Load Factor, g's	3.75*	3.75*	3.75*	3.75*
Surface Areas, ft^2				
Wing, S	2,075	2,075	963	1,300
Horiz. Tail, S_h	588	588	307	399
Vert. Tail, S_v	356	356	124	212
Empenn. Area, S_{emp}	944	944	431	611

*Assumed

Table A8.1a Group Weight Data for Military Trainers

Type	Northrop T-38A	Rockwell NAA T-39A	Cessna T-37A	Fouga Magister	Canadair CL-41
Number of engines:	2	2	2	2	2
Weight Item, lbs					
Wing Group	765	1,753	531	1,089	892
Empennage Group	305	297	128	165	201
Fuselage Group	1,985	2,014	839	743	955
Engine Section	147	315*		in fuse.	40
Land. Gear Group	457	728	330	459	318
Nose Gear					
Main Gear					
Structure Total	3,659	5,107	1,828	2,456	2,406
Engine(s)	1,038	959	751		
Air Induct. System	136	12	14		
Fuel System	285	190	225		
Propulsion System	171	140	205		
Power Plant Total	1,630	1,301	1,195		
Avionics + Instrum.	211	122	132		
Surface Controls	425	344	154		172
Hydraulic System Pneumatic System	154	116	56		
Electrical System	296	720	194		
Electronics	246	407	86		
Air Cond. System** Anti-icing System	142	333	69		
Furnishings	460	857	256		
Auxiliary Gear	24		3		
Fixed Equipm't Total	1,958	2,899	950		
$W_{oil} + W_{tfo}$	62	89	104		
Max. Fuel Capacity	3,916	5,805	1,959	1,299	2,082
Payload (Max. Fuel)***	426	1,500	400	400	400

*Nacelle group for T-39A
**Includes pressurization system
***Includes crew

Table A8.1b Group Weight Data for Military Trainers

Type	Northrop T-38A	Rockwell NAA T-39A	Cessna T-37A	Fouga Magister	Canadair CL-41
Flight Design Gross Weight, GW, lbs	11,651	16,316	6,228	6,280	11,288
Structure/GW	0.314	0.313	0.294	0.391	0.213
Power Plant/GW	0.140	0.080	0.192		
Fixed Equipm't/GW	0.168	0.178	0.152	N.A.	N.A.
Empty Wt/GW	0.622	0.570	0.638	0.755	0.576
Wing Group/GW	0.066	0.107	0.085	0.173	0.079
Emp. Group/GW	0.026	0.018	0.021	0.026	0.018
Fusel.Group/GW	0.170	0.123	0.135	0.118	0.085
Engine Section/GW	0.013	0.019*		in fus.	0.004
Land. Gear Group/GW	0.039	0.045	0.053	0.073	0.028
Take-off Gross Wht, W_{TO}, lbs	11,651	16,701	6,436	6,280	11,288
Empty Weight, W_E, lbs	7,247	9,307	3,973	4,740	5,296
Wing Group/S, psf	4.5	5.1	3.9	5.9	4.1
Emp. Grp/S_{emp}, psf	2.9	2.5	1.7	3.4	3.4
Ultimate Load Factor, g's	10.0**	6.0	10.0		
Surface Areas, ft^2					
Wing, S	170	342	135	186	220
Horiz. Tail, S_h	59	77	54	***	41.3
Vert. Tail, S_v	47.8	41.6	20.4	***	17.5
Empenn. Area, S_{emp}	107	119	74.4	48.8	58.8

*Nacelle group for T-39A
**Assumed
***V-tail

Table A9.1a Group Weight Data for Fighters (USAF)

Type	NAA	McDonnell		Gen. Dyn.	
	F-100F	F-101B	RF-101C	F-102A*	F-16
Number of Engines:	1	2	2	1	1
Weight Item, lbs					
Wing Group	3,896	3,507	3,680	3,000	1,699
Empennage Group	979	812	837	535	650
Fuselage Group	4,032	3,901	3,955	3,409	3,069
Engine Section	104	99	103	39	598
Land. Gear Group	1,509	1,592	1,596	1,056	867
Nose Gear					
Main Gear					
Structure Total	10,520	9,911	10,171	8,039	6,883
Engine(s)	5,121	10,800	9,676	4,993	3,019
Air Induct. System	504	729	638	693	***
Fuel System	761	1,226	1,412	394	349
Propulsion System	414	892	599	278	283
Power Plant Total	6,800	13,647	12,325	6,358	3,651
Avionics + Instrum.	303	318	204	141	994+179
Surface Controls	1,076	772	780	413	719
Hydraulic System	157	433	359	318	361
Pneumatic System					
Electrical System	568	825	819	594	380
Electronics	496	2,222	629	2,001	
Armament	794	228	36	589	566
Air Cond. System**	435	270	362	259	233
Anti-icing System					
Furnishings	427	480	242	227	594
Auxiliary Gear	77	84	91	78	
Auxiliary Power Unit					165
Fixed Equipment Total	4,333	5,632	3,522	4,620	4,191
$W_{oil} + W_{tfo}$	166	223	223	216	
Max. Fuel Capacity	7,729	8,892	9,782	7,053	
Payload (Max. Fuel)	250	1,881	704	1,241	

* This airplane is a delta wing configuration
** Includes pressurization system
*** Air induction system weight of G.D. F-16 is included in fuselage group weight

Table A9.1b Group Weight Data for Fighters (USAF)

Type	NAA F-100F	McDonnell F-101B	McDonnell RF-101C	Gen.Dyn. F-102A*
Flight Design Gross Weight, GW, lbs	29,391	39,800	37,000	25,500
Structure/GW	0.358	0.249	0.275	0.315
Power Plant/GW	0.231	0.343	0.333	0.249
Fixed Equipm't/GW	0.147	0.142	0.095	0.181
Empty Weight/GW	0.737	0.733	0.724	0.750
Wing Group/GW	0.133	0.088	0.099	0.118
Empenn. Group/GW	0.033	0.020	0.023	0.021
Fuselage Group/GW	0.137	0.098	0.107	0.134
Engine Section/GW	0.004	0.002	0.003	0.002
Land. Gear Group/GW	0.051	0.040	0.043	0.041
Take-off Gross Wht, W_{TO}, lbs	30,638	41,288	37,723	28,137
Empty Weight, W_E, lbs	21,653	29,190	26,774	19,130
Wing Group/S, psf	9.7	9.5	10.0	4.3
Emp. Grp/S_{emp}, psf	6.3	5.1	5.2	5.6
Ultimate Load Factor, g's	7.33	10.2	11.0	10.5
Surface Areas, ft^2				
Wing, S	400	368	368	698
Horiz. Tail, S_h	98.9	75.1	75.1	0
Vert. Tail, S_v	55.6	84.9	84.9	95.1
Empenn. Area, S_{emp}	155	160	160	95.1

* This airplane is a delta wing configuration

Table A9.2a Group Weight Data for Fighters (USAF)

Type	Republic F-105B	Gen.Dyn. F-106A*	North American F-107A	F-86H
Number of engines:	1	1	1	1
Weight Item, lbs				
Wing Group	3,409	3,302	3,787	2,702
Empennage Group	965	693	1,130	329
Fuselage Group	5,870	4,401	4,792	2,035
Engine Section	106	39	260	42
Land. Gear Group	1,848	1,232	1,410	989
Nose Gear				
Main Gear				
Structure Total	12,198	9,667	11,379	6,097
Engine	6,187	5,816	6,100	3,646
Air Induct. System	524	975	833	167
Fuel System	608	777	983	845
Propulsion System	406	503	368	340
Power Plant Total	7,725	8,071	8,284	4,998
Avionics + Instrum.	227	190	288	111
Surface Controls	1,311	445	1,454	358
Hydraulic System Pneumatic System	449	431	150	339
Electrical System	700	606	447	476
Electronics	737	2,743	382	230
Armament	719	626	1,006	828
Air Cond. System** Anti-icing System	168	407	390	205
Furnishings	243	290	282	182
Auxiliary Gear	92	69	42	12
APU	224	0	0	0
Fixed Equipm't Total	4,870	5,807	4,441	2,741
$W_{oil} + W_{tfo}$	198	303	143	57
Max. Fuel Capacity	7,540	8,476	11,050	3,660
Payload (Max. Fuel)	757	1,374	2,560	420

*This airplane is a delta wing configuration
**Includes pressurization system

Table A9.2b Group Weight Data for Fighters (USAF)

Type	Republic F-105B	Gen.Dyn. F-106A*	North American F-107A	F-86H
Flight Design Gross Weight, GW, lbs	31,392	30,590	29,524	19,012
Structure/GW	0.389	0.316	0.385	0.321
Power Plant/GW	0.246	0.264	0.281	0.263
Fixed Equipm't/GW	0.155	0.190	0.150	0.144
Empty Weight/GW	0.797	0.766	0.816	0.728
Wing Group/GW	0.109	0.108	0.128	0.142
Empenn. Group/GW	0.031	0.023	0.038	0.017
Fuselage Group/GW	0.187	0.144	0.162	0.107
Engine Section/GW	0.003	0.001	0.009	0.002
Land. Gear Group/GW	0.059	0.040	0.048	0.052
Take-off Gross Wht, W_{TO}, lbs	34,081	33,888	39,405	18,908
Empty Weight, W_E, lbs	25,022	23,448	24,104	13,836
Wing Group/S, psf	8.9	4.7	9.6	8.6
Emp. Grp/S_{emp}, psf	5.2	6.6	6.4	4.1
Ultimate Load Factor, g's	13.0	10.5	13.0	11.0
Surface Areas, ft^2				
Wing, S	385	698	395	313
Horiz. Tail, S_h	96.5	0	93.3	47.2
Vert. Tail, S_v	88.1	105	83.8	32.2
Empenn. Area, S_{emp}	185	105	177	79.4

* This airplane is a delta wing configuration

Table A9.3a Group Weight Data for Fighters (USN)

Type	Vought F8U-3	McDonnell F4H	Grumman F11F	F9F-5
Number of engines:	1	2	1	1
Weight Item, lbs				
Wing Group	4,128	4,343	2,180	2,294
Empennage Group	1,045	853	669	404
Fuselage Group	3,850	4,042	3,269	1,779
Engine Section	92	125	47	0
Land. Gear Group	949	1,735	907	728
Nose Gear				
Main Gear				
Structure Total	10,064	11,098	7,072	5,205
Engine(s)	6,010	6,940	3,489	2,008
Air Induct. System	673	1,037	159	225
Fuel System	849	953	463	529
Propulsion System.	338	106	192	116
Power Plant Total	7,870	9,036	4,303	2,878
Avionics + Instrum.	191	166	118	82
Surface Controls	1,425	919	760	345
Hydraulic System Pneumatic System	150	441	166	267
Electrical System	439	502	459	458
Electronics	840	1,386	439	292
Armament	376	446	358	416
Air Cond. System* Anti-icing System	329	341	76	85
Furnishings	210	321	166	144
Auxiliary Gear	183	0	131	51
Fixed Equipm't Total	4,143	4,522	2,673	2,140
$W_{oil} + W_{tfo}$	196	131	72	
Max. Fuel Capacity	14,306	13,410	6,663	7,160
Payload (Max. Fuel)	1,197	1,500	340	

*Includes pressurization system

Table A9.3b Group Weight Data for Fighters (USN)

Type	Vought F8U-3	McDonnell F4H	Grumman F11F	F9F-5
Flight Design Gross Weight, GW, lbs	30,578	34,851	17,500	14,900
Structure/GW	0.329	0.318	0.404	0.349
Power Plant/GW	0.257	0.259	0.246	0.193
Fixed Equipm't/GW	0.135	0.130	0.153	0.144
Empty Weight/GW	0.722	0.707	0.771	0.686
Wing Group/GW	0.135	0.125	0.125	0.154
Empenn. Group/GW	0.034	0.024	0.038	0.027
Fuselage Group/GW	0.126	0.116	0.187	0.119
Engine Section/GW	0.003	0.004	0.003	0
Land. Gear Group/GW	0.031	0.050	0.052	0.049
Take-off Gross Wht, W_{TO}, lbs	38,528	40,217	21,233	17,500
Empty Weight, W_E, lbs	22,092	24,656	13,485	10,223
Wing Group/S, psf	8.9	8.2	8.5	9.2
Emp. Grp/S_{emp}, psf	7.2	5.2	5.8	3.5
Ultimate Load Factor, g's	9.6		9.8	11.3
Surface Areas, ft^2				
Wing, S	462	530	255	250
Horiz. Tail, S_h	67.2	96.2	65.5	48
Vert. Tail, S_v	78.6	67.5	50.3	66
Empenn. Area, S_{emp}	146	164	116	114

Table A9.4a Group Weight Data for Fighters (USN)

Type	Grumman A2F(A6)	McDonnell F3H-2	NAA A3J	Vought F7U-1
Number of engines:	2	2	2	1
Weight Item, lbs				
Wing Group	4,733	4,314	5,072	3,583
Empennage Group	819	576	1,358	726
Fuselage Group	3,538	3,551	6,851	937
Engine Section	64	93	80	
Land. Gear Group	2,343	1,458	2,173	1,181
Nose Gear				
Main Gear				
Structure Total	11,497	9,992	15,534	6,427
Engine(s)	4,010	4,960	7,260	2,790
Air Induct. System	61	614	767	690
Fuel System	936	1,262	979	1,080
Propulsion System.	632	70	353	937
Power Plant Total	5,639	6,906	9,359	5,497
Avionics + Instrum.	133	145	210	108
Surface Controls	932	1,067	1,845	482
Hydraulic System Pneumatic System	170	474	275	317
Electrical System	695	535	821	371
Electronics	2,652	984	2,239	328
Armament	323	662	45	367
Air Cond. System* Anti-icing System	164	101	424	79
Furnishings	476	218	676	279
Auxiliary Gear		253		128
Fixed Equipm't Total	5,545	4,439	6,535	2,459
$W_{oil} + W_{tfo}$	195	147	320	97
Max. Fuel Capacity	8,764	9,789	19,074	5,826
Payload (Max. Fuel)	2,000	216	1,885	502

*Includes pressurization system

Table A9.4b Group Weight Data for Fighters (USN)

Type	Grumman A2F(A6)	McDonnell F3H-2	NAA A3J	Vought F7U-1*
Flight Design Gross Weight, GW, lbs	34,815	26,000	46,028	19,310
Structure/GW	0.330	0.384	0.337	0.333
Power Plant/GW	0.162	0.266	0.203	0.285
Fixed Equipm't/GW	0.159	0.171	0.142	0.127
Empty Weight/GW	0.651	0.818	0.679	0.746
Wing Group/GW	0.136	0.166	0.110	0.186
Empenn. Group/GW	0.024	0.022	0.030	0.038
Fuselage Group/GW	0.102	0.137	0.149	0.048
Engine Section/GW	0.002	0.004	0.002	
Land. Gear Group/GW	0.067	0.056	0.047	0.061
Take-off Gross Wht, W_{TO}, lbs	34,815	32,037	53,658	21,638
Empty Weight, W_E, lbs	22,680	21,272	31,246	14,397
Wing Group/S, psf	9.1	8.4	7.2	7.1
Emp. Grp/S_{emp}, psf	4.4	4.5	3.4	8.3
Ultimate Load Factor, g's		11.25		
Surface Areas, ft^2				
Wing, S	520	516	700	507
Horiz. Tail, S_h	120	82.5	304	0*
Vert. Tail, S_v	68.4	45.4	101	88
Empenn. Area, S_{emp}	188	128	405	88

*This airplane is essentially a flying wing with two vertical tails

Table A9.5a Group Weight Data for Fighters (USAF and USN)

Type	McDonnell Douglas F-4E (USAF)	F-15C (USAF)	F/A-18A (USN)	AV-8B* (USN)
Number of engines:	2	2	2	1
Weight Item, lbs				
Wing Group	5,226	3,642	3,798	1,443
Empennage Group	969	1,104	945	372
Fuselage Group	5,050	6,245	4,685	2,060
Engine Section	166	102	143	141
Land. Gear Group	1,944	1,393	1,992	1,011
Nose Gear	377	264	626	334
Main Gear	1,567	1,129	1,366	400
Outrigger Gear				277
Structure Total	13,355	12,486	11,563	5,027
Engine(s)	7,697	6,091	4,294	3,815
Air Induct. System	1,318	1,464	423	236
Fuel System	1,932	1,128	1,002	542
Propulsion System.	312	522	558	444
Power Plant Total	11,259	9,205	6,277	5,037
Instrument group	270	151	94	80
Surface Controls	1,167	810	1,067	698
Hydraulic System Pneumatic System	543	433	364	176
Electrical System	542	607	547	424
Electronics	2,227	1,787	1,538	697
Armament	641	627	387	152
Air Cond. System Pressurization Syst.	406	685	593	218
Anti-icing System			21	
Furnishings	611	294	317	298
Auxiliary Gear	412	119	189	
Photographic System		24		
Ballast		318	36	
Manuf. Variation	57	-97	-19	-16
Fixed Equipm't Total	6,900	5,734	5,134	2,727
Max. Fuel Capacity	12,058	13,455	17,592**	7,759
Expendable Payload	2,193	2,571	5,453	4,271
Fixed Payload***	incl. in armament		2,231	832

*V/STOL fighter **Incl. 6,732 lbs ext. fuel ***Pylons, racks, launchers, FLIR and camera pods

Table A9.5b Group Weight Data for Fighters (USAF and USN)

Type	McDonnell Douglas F-4E	F-15C	F/A-18A	AV-8B*
Flight Design Gross Weight, GW, lbs	37,500	37,400	32,357	22,950
Structure/GW	0.356	0.334	0.357	0.219
Power Plant/GW	0.300	0.246	0.194	0.219
Fixed Equipm't/GW	0.182	0.147	0.158	0.120
Empty Weight/GW	0.840	0.733	0.710	0.557
Wing Group/GW	0.139	0.097	0.117	0.063
Empenn. Group/GW	0.026	0.030	0.029	0.016
Fuselage Group/GW	0.135	0.167	0.145	0.090
Engine Section/GW	0.004	0.003	0.004	0.006
Land. Gear Group/GW	0.052	0.037	0.062	0.044
Take-off Gross Wht, W_{TO}, lbs	58,000	68,000	51,900	29,750
Empty Weight, W_E, lbs	31,514	27,425	22,974	12,791
Wing Group/S, psf	9.5	6.1	9.5	6.3
Emp. Grp/S_{emp}, psf	5.8	4.7	4.9	5.0
Ultimate Load Factor, g's	9.75	11.0	11.25	10.5
Surface Areas, ft^2				
Wing, S	548	599	400	230
Horiz. Tail, S_h	100	111	88.1	48.5
Vert. Tail, S_v	67.5	125	104	26.6
Empenn. Area, S_{emp}	168	236	192	75.1

*V/STOL Fighter

Table A10.1a Group Weight Data for Military Jet Transports

Type	Boeing KC135*	Lockheed C-141B	C-5A
Number of engines:	4	4	4
Weight Item, lbs			
Wing Group	25,251	35,272	100,015
Empennage Group	5,074	5,907	12,461
Fuselage Group	18,867	36,857	118,193
Nacelle Group	2,575	5,168	9,528
Land. Gear Group	10,180	10,850	38,353
Nose Gear		1,234	4,455
Main Gear		9,616	33,898
Structure Total	61,947	94,054	278,550
Engine(s)	16,687	23,665	38,035
Air Induct. System	172		
Fuel System	4,052	1,802	2,540
Propulsion System	591		
Power Plant Total	21,502	25,467	40,575
Avionics + Instrum.	553	3,078	3,823
Surface Controls	2,044	3,701	7,404
Hydraulic System	858	1,604	4,086
Pneumatic System			
Electrical System	2,470	2,826	3,503
Electronics	2,096	1,163	992
APU	0	534	987
Oxygen System		479	308
Air Cond. System**	1,464	2,283	3,416
Anti-icing System		453	229
Furnishings	1,518	5,210	19,272
Auxiliary Gear	1,899	103	39
Fixed Equipm't Total	12,902	21,434	44,059
$W_{oil} + W_{tfo}$	1,407	1,327	826
Max. Fuel Capacity	158,997	153,352	332,500
Payload (Max.)		73,873	200,000

*This is a tanker airplane
**Includes pressurization system

Table A10.1b Group Weight Data for Military Jet Transports

Type	Boeing KC135*	Lockheed C-141B	C-5A
Flight Design Gross Weight, GW, lbs	297,000	314,200	769,000
Structure/GW	0.209	0.299	0.362
Power Plant/GW	0.072	0.081	0.053
Fixed Equipm't/GW	0.043	0.068	0.057
Empty Weight/GW	0.323	0.449	0.472
Wing Group/GW	0.085	0.112	0.130
Empenn. Group/GW	0.017	0.019	0.016
Fuselage Group/GW	0.064	0.117	0.154
Nacelle Group/GW	0.009	0.016	0.012
Land. Gear Group/GW	0.034	0.035	0.050
Take-off Gross Wht, W_{TO}, lbs	297,000	314,000	769,000
Empty Weight, W_E, lbs	95,938	140,955	363,184
Wing Group/S, psf	10.4	10.9	16.1
Emp. Grp/S_{emp}, psf	6.2	6.6	6.5
Ultimate Load Factor, g's	3.75	3.75	3.75**
Surface Areas, ft^2			
Wing, S	2,435	3,228	6,200
Horiz. Tail, S_h	500	483	966
Vert. Tail, S_v	312	416	961
Empenn. Area, S_{emp}	812	899	1,927

*This is a tanker airplane
**after 100,000 lbs of fuel has been used.

Table A10.2a Group Weight Data for Turbo/Propeller Driven Military Transports

	A.W.(HS) Argosy	Douglas C-133A	Lockheed C-130H	Breguet 941*
Number of engines:	4	4	4	4
Weight Item, lbs				
Wing Group	10,800	27,403	13,950	4,096
Empennage Group	1,300	6,011	3,480	1,387
Fuselage Group	11,100**	30,940	14,695	6,481
Nacelle Group	1,200	3,512	2,756	in wing
Land. Gear Group	3,180	10,635	5,309	2,626
Nose Gear			730	
Main Gear			4,579	
Structure Total	27,580	78,501	40,190	14,590
Engines		10,470	13,746	
Air Induct. System				
Fuel System		1,338	3,105	
Propeller Inst.		5,403	in eng.	
Propulsion System		2,081	in eng.	
Power Plant Total		19,292	16,851	
Avionics + Instrum.		578	3,582	
Surface Controls	in struct.	1,804	1,673	1,056
Hydraulic System		2,678	664	
Pneumatic System				
Electrical System		2,004	2,459	
Electronics		2,047	in avionics	
APU		188	651	
Oxygen System			231	
Air Cond. System***		2,973	1,684	
Anti-icing System			797	
Furnishings		3,632	4,472	
Auxiliary Gear		117	6	
Operating items			532	
Fixed Equipm't Total		16,021	16,219	
$W_{oil} + W_{tfo}$		1,693	1,089	
Max. Fuel Capacity		60,000	45,240	
Payload (Max.)		97,162	33,461	

*This is a STOL airplane **Tailbooms at 2,360 lbs are included ***Includes pressurization system

Table A10.2b Group Weight Data for Turbo/Propeller Driven Military Transports

Type	A.W.(HS) Argosy	Douglas C-133A	Lockheed C-130H	Breguet 941
Flight Design Gross Weight, GW, lbs	82,000	275,000	155,000	58,421
Structure/GW	0.336	0.285	0.259	0.250
Power Plant/GW		0.070	0.109	
Fixed Equipm't/GW		0.058	0.105	
Empty Weight/GW	0.561	0.414	0.473	0.508
Wing Group/GW	0.132	0.100	0.090	0.070
Empenn. Group/GW	0.016	0.022	0.022	0.024
Fuselage Group/GW	0.135	0.113	0.095	0.111
Nacelle Group/GW	0.015	0.013	0.018	in wing
Land. Gear Group/GW	0.039	0.039	0.034	0.045
Take-off Gross Wht, W_{TO}, lbs	82,000	275,000	155,000	58,421
Empty Weight, W_E, lbs	46,000	113,814	73,260	29,675
Wing Group/S, psf	7.4	10.3	8.0	4.5
Emp. Grp/S_{emp}, psf	2.3	4.2	4.2	2.6
Ultimate Load Factor, g's	3.75*	2.50	3.75*	3.75*
Surface Areas, ft^2				
Wing, S	1,458	2,673	1,745	902
Horiz. Tail, S_h	327	801	536	320
Vert. Tail, S_v	250	641	300	223
Empenn. Area, S_{emp}	577	1,442	836	543

*Assumed

Table A10.3a Group Weight Data for Piston/Propeller Driven Military Transports

<table>
<tr><th>Type</th><th>Beech
L-23F*</th><th>Chase
C-123B</th><th>DeHavill.
Caribou</th><th>Fairchild
C-119B</th></tr>
<tr><td>Number of engines:</td><td>2</td><td>2</td><td>2</td><td>2</td></tr>
<tr><td>Weight Item, lbs</td><td></td><td></td><td></td><td></td></tr>
<tr><td>Wing Group</td><td>692</td><td>6,153</td><td>2,925</td><td>7,226</td></tr>
<tr><td>Empennage Group</td><td>156</td><td>1,103</td><td>790</td><td>1,193</td></tr>
<tr><td>Fuselage Group</td><td>679</td><td>7,763</td><td>2,849</td><td>7,157</td></tr>
<tr><td>Nacelle Group</td><td>279</td><td>1,633</td><td>781</td><td>2,538**</td></tr>
<tr><td>Land. Gear Group</td><td>453</td><td>2,081</td><td>1,230</td><td>4,197</td></tr>
<tr><td>Nose Gear</td><td></td><td></td><td></td><td></td></tr>
<tr><td>Main Gear</td><td></td><td></td><td></td><td></td></tr>
<tr><td>Structure Total</td><td>2,259</td><td>18,733</td><td>8,575</td><td>22,311</td></tr>
<tr><td>Engines</td><td>1,015</td><td>4,810</td><td>3,170</td><td>6,500</td></tr>
<tr><td>Propellers</td><td>260</td><td>1,430</td><td>871</td><td></td></tr>
<tr><td>Fuel System</td><td>127</td><td>671</td><td>221</td><td></td></tr>
<tr><td>Propulsion System</td><td>215</td><td>1,103</td><td>453</td><td></td></tr>
<tr><td>Power Plant Total</td><td>1,617</td><td>8,014</td><td>4,715</td><td>11,979</td></tr>
<tr><td>Avionics + Instrum.</td><td>91</td><td>161</td><td>121</td><td></td></tr>
<tr><td>Surface Controls</td><td>129</td><td>490</td><td>326</td><td></td></tr>
<tr><td>Hydraulic System</td><td>0</td><td rowspan="2">148</td><td rowspan="2">85</td><td></td></tr>
<tr><td>Pneumatic System</td><td>0</td><td></td></tr>
<tr><td>Electrical System</td><td>190</td><td>904</td><td>436</td><td></td></tr>
<tr><td>Electronics</td><td>80</td><td>452</td><td>273</td><td></td></tr>
<tr><td>APU</td><td>0</td><td>136</td><td>0</td><td></td></tr>
<tr><td>Air Cond. System</td><td rowspan="2">104</td><td rowspan="2">642</td><td rowspan="2">82</td><td></td></tr>
<tr><td>Anti-icing System</td><td></td></tr>
<tr><td>Furnishings</td><td>459</td><td>428</td><td>271</td><td></td></tr>
<tr><td>Auxiliary Gear</td><td>7</td><td>0</td><td>16</td><td></td></tr>
<tr><td>Fixed Equipm't Total</td><td>1,060</td><td>3,361</td><td>1,610</td><td>6,834</td></tr>
<tr><td>$W_{oil} + W_{tfo}$</td><td>92</td><td>420</td><td>406</td><td></td></tr>
<tr><td>Water</td><td>0</td><td>225</td><td>0</td><td></td></tr>
<tr><td>Max. Fuel Capacity</td><td>1,080</td><td>5,452</td><td></td><td>15,540</td></tr>
<tr><td>Payload (Max.)</td><td>1,090</td><td>16,000</td><td>7,344</td><td></td></tr>
</table>

*Military version of Twin Bonanza
**Tailbooms included

Table A10.3b Group Weight Data for Piston/Propeller Driven Military Transports

Type	Beech L-23F*	Chase C-123B	DeHavill. Caribou	Fairchild C-119B
Flight Design Gross Weight, GW, lbs	7,368	54,000	26,000	64,000
Structure/GW	0.307	0.347	0.330	0.349
Power Plant/GW	0.219	0.148	0.181	0.187
Fixed Equipm't/GW	0.144	0.062	0.062	0.107
Empty Weight/GW*	0.670	0.558	0.635	0.641
Wing Group/GW	0.094	0.114	0.113	0.113
Empenn. Group/GW	0.021	0.020	0.030	0.019
Fuselage Group/GW	0.092	0.144	0.110	0.112
Nacelle Group/GW	0.038	0.030	0.030	0.040**
Land. Gear Group/GW	0.061	0.039	0.047	0.066
Take-off Gross Wht, W_{TO}, lbs	7,368	52,802	26,000	64,000
Empty Weight, W_E, lbs	4,936	30,108	16,500	41,017
Wing Group/S, psf	2.5	5.0	3.2	5.0
Emp. Grp/S_{emp}, psf	1.7	2.1	1.9	2.7
Ultimate Load Factor, g's	6.6	4.5	5.1	
Surface Areas, ft^2				
Wing, S	277	1,223	912	1,447
Horiz. Tail, S_h	29.3		206	297
Vert. Tail, S_v	65.1		211	151
Empenn. Area, S_{emp}	94.4	520	417	448

*This is a military version of the Twin Bonanza
**Tailbooms included

Table A10.4a Group Weight Data for Piston/Propeller Driven Military Transports

Type	Douglas C-124C	Boeing C-97C	Lockheed C-69	C-121A
Number of engines:	4	4	4	4
Weight Item, lbs				
Wing Group	18,135	15,389	9,466	11,184
Empennage Group	3,025	2,078	2,026	2,094
Fuselage Group	18,073	13,572	6,794	8,520
Nacelle Group	6,119	4,753	2,505	3,970
Land. Gear Group	11,701	7,112	4,481	4,771
Nose Gear		963	1,019	1,077
Main Gear		6,149	3,462	3,694
Structure Total	57,053	42,904	25,272	30,539
Engine(s)	15,551	13,844	10,568	11,536
Air Induct. System	4,046			
Fuel System	4,059			
Propeller Inst.	4,363			
Propulsion System	in prop.			
Power Plant Total	28,019	23,051	15,633	15,676
Avionics + Instrum.	769			
Surface Controls	1,493			
Hydraulic System Pneumatic System	582			
Electrical System	1,952			
Electronics	1,886			
APU	410			
Air Cond. System* Anti-icing System	3,294			
Furnishings	7,539			
Auxiliary Gear	104			
Fixed Equipm't Total	18,029	9,997	7,625	13,710
$W_{oil} + W_{tfo}$	3,389			
Water and Alcohol	522			
Max. Fuel Capacity	22,000	23,094	33,470	41,496
Payload (Max.)	55,262	46,500	12,330	12,550

*Includes pressurization system

Table A10.4b Group Weight Data for Piston/Propeller Driven Military Transports

Type	Douglas C-124C	Boeing C-97C	Lockheed C-69	C-121A
Flight Design Gross Weight, GW, lbs	185,000	150,000	82,000	132,800
Structure/GW	0.308	0.286	0.308	0.230
Power Plant/GW	0.151	0.154	0.191	0.118
Fixed Equipm't/GW	0.097	0.067	0.093	0.103
Empty Weight/GW	0.552	0.506	0.592	0.450
Wing Group/GW	0.098	0.103	0.115	0.084
Empenn. Group/GW	0.016	0.014	0.025	0.016
Fuselage Group/GW	0.098	0.090	0.083	0.064
Nacelle Group/GW	0.033	0.032	0.031	0.030
Land. Gear Group/GW	0.063	0.047	0.055	0.036
Take-off Gross Wht, W_{TO}, lbs	185,000	150,000	82,000	132,800
Empty Weight, W_E, lbs	102,181	75,974	48,530	59,715
Wing Group/S, psf	7.2	8.7	5.7	6.8
Emp. Grp/S_{emp}, psf	2.6	3.3	2.9	3.0
Ultimate Load Factor, g's	3.75	3.75	3.75	3.75
Surface Areas, ft^2				
Wing, S	2,506	1,769	1,650	1,650
Horiz. Tail, S_h	681	333	464	464
Vert. Tail, S_v	465	306	262	242
Empenn. Area, S_{emp}	1,146	639	706	706

Table A10.5a Group Weight Data for Military Patrol Airplanes

Type	Grumman S2F-1	Lockheed P2V-4	Lockheed U2
Number of engines:	2	2	1
Weight Item, lbs	Piston/Propeller		Jet
Wing Group	2,902	7,498	2,034
Empennage Group	681	1,589	320
Fuselage Group	1,701	5,155	1,410
Nacelle Group	965	2,303	0
Land. Gear Group	1,396	3,715	263
Nose Gear			tail gear 60
Main Gear			203
Structure Total	7,645	20,260	4,027
Engines	2,953	5,726	4,076
Propellers	866	1,137	
Fuel System	215	2,827	311
Propulsion System	390	1,633	479**
Power Plant Total	4,424	11,323	4,866
Avionics + Instrum.	147	194	57
Surface Controls	714	960	362
Hydraulic System / Pneumatic System	208	284	66
Electrical System	988	1,503	290
Electronics	2,310	2,903	166
Armament	256	1,705	
Air Cond. System* / Anti-icing System	356	496	135
Furnishings	657	1,327	82
Auxiliary Gear	281	0	193
Fixed Equipm't Total	5,917	9,372	1,351
$W_{oil} + W_{tfo}$	332	1,640	97
Engine oil			120
Armament Provisions	18	5,951	
Water	0	1,480	
Max. Fuel Capacity	3,126	14,006	5,810
Payload (Max.)	1,938		518
Crew	N.A.	N.A.	285

*Incl. press. system **Incl. air induction and exhausts

Table A10.5b Group Weight Data for Military Patrol Airplanes

Type	Grumman S2F-1 Piston/Propeller	Lockheed P2V-4 Piston/Propeller	Lockheed U2 Jet
Flight Design Gross Weight, GW, lbs	23,180	67,500	17,000
Structure/GW	0.330	0.300	0.237
Power Plant/GW	0.191	0.168	0.286
Fixed Equipm't/GW	0.255	0.139	0.079
Empty Weight/GW	0.775	0.607	0.603
Wing Group/GW	0.125	0.111	0.120
Empenn. Group/GW	0.029	0.024	0.019
Fuselage Group/GW	0.073	0.076	0.083
Nacelle Group/GW	0.042	0.034	
Land. Gear Group/GW	0.060	0.055	0.015
Take-off Gross Wht, W_{TO}, lbs	24,167	67,500	19,913
Empty Weight, W_E, lbs	17,953	40,955	10,244
Wing Group/S, psf	6.0	7.5	3.4
Emp. Grp/S_{emp}, psf	3.5	3.9	2.3
Ultimate Load Factor, g's	4.5	4.0	3.75
Surface Areas, ft^2			
Wing, S	485	1,000	600
Horiz. Tail, S_h	103	241	90
Vert. Tail, S_v	90.2	170	49
Empenn. Area, S_{emp}	193	411	139

Table A11.1a Group Weight Data for Flying Boats, Amphibious and Float Airplanes

Type	At the time of printing no data were available
Number of engines:	
Weight Item, lbs	
Wing Group	
Empennage Group	
Fuselage Group	
Nacelle Group	
Land. Gear Group	
Nose Gear	
Main Gear	

Structure Total	

Engines	
Propellers	
Fuel System	
Propulsion System	

Power Plant Total	

Avionics + Instrum.	
Surface Controls	
Hydraulic System	
Pneumatic System	
Electrical System	
Electronics	
Armament	
Air Cond. System	
Anti-icing System	
Furnishings	
Auxiliary Gear	

Fixed Equipm't Total	

$W_{oil} + W_{tfo}$	
Armament Provisions	
Water	
Max. Fuel Capacity	
Payload (Max.)	

Table A11.1b Group Weight Data for Flying Boats, Amphibious and Float Airplanes

Type	At the time of printing no data were available
Flight Design Gross Weight, GW, lbs	
Structure/GW	
Power Plant/GW	
Fixed Equipm't/GW	
Empty Weight/GW	
Wing Group/GW	
Empenn. Group/GW	
Fuselage Group/GW	
Nacelle Group/GW	
Land. Gear Group/GW	
Take-off Gross Wht, W_{TO}, lbs	
Empty Weight, W_E, lbs	
Wing Group/S, psf	
Emp. Grp/S_{emp}, psf	
Ultimate Load Factor, g's	
Surface Areas, ft^2	
Wing, S	
Horiz. Tail, S_h	
Vert. Tail, S_v	
Empenn. Area, S_{emp}	

Table A12.1a Group Weight Data for Supersonic Cruise Airplanes

Type	AST-100 *	SSXJET **	Super-cruiser ***
Number of engines:	4	2	2
Weight Item, lbs			
Wing Group	85,914	3,599	3,962
Empennage Group	10,655	481	225
Fuselage Group	52,410	3,494	2,195
Nacelle Group	16,803	505	700
Land. Gear Group	27,293	1,391	1,300
Nose Gear			
Main Gear			
Structure Total	193,075	9,470	8,382
Engines	52,000	3,016	4,781
Air Induct. System	incl. in propulsion system		
Fuel System	5,781	626	560
Propulsion System	1,780	59	165
Power Plant Total	59,561	3,701	6,756
Avionics + Instrum.	6,090	660	1,484
Surface Controls	9,405	564	1,207
Hydraulic System	5,600	266	
Pneumatic System			302
Electrical System	5,050	445	357
Electronics	incl. in avionics + instrum.		
Armament	0	0	600
Air Cond. System****	8,200	352	
Anti-icing System	210	95	250
Furnishings	25,111	417	242
Auxiliary Gear	0	0	40
Fixed Equipm't Total	59,666	2,799	4,482
$W_{oil} + W_{tfo}$	3,050	133	
Mission Fuel Reqd.	327,493	18,674	12,523
Payload	61,028	725	5,000

*NASA TM X-73936, M=2.2 large passenger transport
**NASA TM 74055, M=2.2 executive (business) jet
***NASA TM 78811, M=2.6 military missile carrying super cruiser
****Includes pressurization system

Table A12.1b Group Weight Data for Supersonic Cruise Airplanes

Type	AST-100 *	SSXJET **	Super-cruiser ***
Flight Design Gross Weight, GW, lbs	718,000	35,720	37,144
Structure/GW	0.269	0.265	0.226
Power Plant/GW	0.083	0.104	0.182
Fixed Equipm't/GW	0.083	0.078	0.121
Empty Weight/GW	0.435	0.447	0.528
Wing Group/GW	0.120	0.101	0.107
Empenn. Group/GW	0.015	0.013	0.006
Fuselage Group/GW	0.073	0.098	0.059
Nacelle Group/GW	0.023	0.014	0.019
Land. Gear Group/GW	0.038	0.039	0.035
Take-off Gross Wht, W_{TO}, lbs	718,000	35,720	47,900
Empty Weight, W_E, lbs	312,302	15,970	19,620
Wing Group/S, psf	8.6	3.7	10.7
Emp. Grp/S_{emp}, psf	11.0	3.9	3.1
Ultimate Load Factor, g's	3.75	3.75	6.0
Surface Areas, ft^2			
Wing, S	9,969	965	371
Horiz. Tail, S_h	579	62	0
Vert. Tail, S_v	386	62	73
Empenn. Area, S_{emp}	965	124	73

*NASA TM X-73936, M=2.2 large passenger transport
**NASA TM 74055, M=2.2 executive (business) jet
***NASA TM 78811, M=2.6 military missile carrying super cruiser

Table A13.1a Group Weight Data for NASA X Airplanes

Type	Ryan X-13*	North American X-15**	Hiller X-18***	Bell XV-15
Number of engines:	1	1	2	2
Weight Item, lbs	Jet	Rocket	TBP	TBP Tiltrotor
Wing Group	515	1,144	3,483	946
Empennage Group	146	1,267	928	259
Fuselage Group	415	3,806	4,694	1,589
Engine Section	69	187	728	nac. 369
Land. Gear Group	300	389	1,289	524
Nose Gear				
Main Gear				
Structure Total	1,445	6,793	11,122	3,687
Engine(s)	2,766	680	5,460	1,052
Fuel System	100	1,354	623	226
Propeller Inst.	0	0	3,679	863
Propulsion System	227	148	548	222
Drive system				1,340
Propeller Controls	0	0	2,023	****629
Power Plant Total	3,093	2,182	12,333	4,332
Avionics + Instrum.	41	172	141	231
Surface Controls	416	1,182	896	****777
Hydraulic System	214	240	863	util. 86
Electrical System	311	142	931	418
Electronics	29	175	63	
Test Instrumentation	139	1,328		1,160
Ballast				106
Air Cond. System	10	192		119
Furnishings	199	446	813	434
Auxiliary Gear	0	11	0	10
Fixed Equipm't Total	1,359	3,888	3,707	3,341
$W_{oil} + W_{tfo}$	43	0	353	32
Liquid Nitrogen		150	engine oil:	53
Max. Fuel Capacity	1,400	314	823	1,401
Payload	crew and test equipment only			

*Delta configuration, took off from vertical position
**Air launched by B-52
***Turbo/propeller driven, wing incidence variable over more than 90 degrees
****hydraulic system incl. in rotor and surface ctrls

Table A13.1b Group Weight Data for NASA X Airplanes

Type	Ryan X-13*	North American X-15**	Hiller X-18***	**** Bell XV-15
Flight Design Gross Weight, GW, lbs	7,000	13,592	33,000	13,226
Structure/GW	0.206	0.500	0.337	0.279
Power Plant/GW	0.442	0.161	0.374	0.328
Fixed Equipm't/GW	0.194	0.286	0.112	0.253
Empty Weight/GW	0.822	0.949	0.826	0.859
Wing Group/GW	0.074	0.084	0.106	0.072
Empenn. Group/GW	0.021	0.093	0.028	0.020
Fuselage Group/GW	0.059	0.280	0.142	0.120
Engine Section/GW	0.010	0.014	0.022	0.028
Land. Gear Group/GW	0.043	0.029	0.039	0.040
Take-off Gross Wht, W_{TO}, lbs	7,149	13,592	33,000	13,226
Empty Weight, W_E, lbs	5,755	12,901	27,272	11,360
Wing Group/S, psf	2.7	10.9	6.6	5.6
Emp. Grp/S_{emp}, psf	2.3	10.0	2.8	2.6
Ultimate Load Factor, g's	6.0	11.0		
Surface Areas, ft^2				
Wing, S	191	105	528	169
Horiz. Tail, S_h	0	52	201	50.3
Vert. Tail, S_v	62.8	74.9	133	50.5
Empenn. Area, S_{emp}	62.8	127	334	101

*Delta configuration, took off from vertical position
**Air launched by B-52
***Turbo/propeller driven, wing incidence variable over more than 90 degrees
****Tiltrotor research airplane

Table A13.2a Group Weight Data for NASA X Airplanes

Type	Bell X-2*	Bell X-5**	Northrop YP-61***	**** Bell XP-77
Number of engines:	1	1	2	1
Weight Item, lbs	Rocket	Jet	Piston/ Prop.	Piston/ Prop.
Wing Group	2,856	1,683	3,969	463
Empennage Group	445	198	629	59
Fuselage Group	4,108	1,064	1,557	218
Engine Section	30	274	1,817 incl.booms	123
Land. Gear Group	421	532	1,803	344
Nose Gear	108	464	303	123
Main Gear	(skids) 313	68	1,500	221
Structure Total	8,281	3,751	9,775	1,207
Engine(s)	607	2,223	4,974	738
Air Induction System		31		
Fuel System	898	108	914	91
Propulsion System	13	77	1,081	101
Propellers			1,111	206
Power Plant Total	1,518	2,439	8,080	1,136
Avionics + Instrum.	65	35	119	37
Surface Controls	364	195	400	42
Hydraulic System	442	139	240	0
Electrical System	604	127	668	92
Electronics	63	86	721	99
Anti-icing System			100	
Test Instrumentation	708	155		
Armament(incl. guns and ammo)			3,364	391
Ballast		98		
Air Cond. System	102	69		
Furnishings	158	84	252	56
Auxiliary Gear	0	90	352	15
Fixed Equipm't Total	2,506	1,078	6,216	732
$W_{oil} + W_{tfo}$	484		152	30
Liquid Oxygen	7,180			
Liquid Nitrogen	26	oil: 23	oil: 270	28
Max. Fuel Capacity	5,716	1,200	3,168	312
Payload	crew and test equipment only			

*Air launched by B50 **Variable sweep wing
***Twin boom fighter ****Wood built lightweight fighter

Table A13.2b Group Weight Data for NASA X Airplanes

Type	Bell X-2*	Bell X-5**	Northrop YP-61***	**** Bell XP-77
Flight Design Gross Weight, GW, lbs	25,627	8,737	27,813	3,632
Structure/GW	0.323	0.429	0.351	0.332
Power Plant/GW	0.059	0.279	0.291	0.313
Fixed Equipm't/GW	0.098	0.123	0.223	0.202
Empty Weight/GW	0.480	0.832	0.865	0.847
Wing Group/GW	0.111	0.193	0.143	0.127
Empenn. Group/GW	0.017	0.023	0.023	0.016
Fuselage Group/GW	0.160	0.122	0.056	0.060
Engine Section/GW	0.001	0.031	0.065 incl.booms	0.034
Land. Gear Group/GW	0.016	0.061	0.065	0.095
Take-off Gross Wht, W_{TO}, lbs	25,627	8,737	27,813	3,632
Empty Weight, W_E, lbs	12,305	7,268	24,071	3,075
Wing Group/S, psf	11.0	9.6	6.0	4.6
Emp. Grp/S_{emp}, psf	4.1	3.4	3.0	2.1
Ultimate Load Factor, g's	not available			
Surface Areas, ft^2				
Wing, S	260	175	664	100
Horiz. Tail, S_h	49.5	31.8	120	18.7
Vert. Tail, S_v	58	25.8	92	9.0
Empenn. Area, S_{emp}	108	57.6	212	27.7

*Air launched by B50 **Variable sweep wing
***Twin boom fighter ****Wood built lightweight fighter

Table A13.3a Group Weight Data for NASA X Airplanes

Type	Mc Donnell XF-88A	Convair XF-92A**	NAA YF-93A***	Convair XFY-1
Number of engines:	2	1	1	1
Weight Item, lbs	Jet	Jet	Jet	TBP****
Wing Group	2,048	1,694	2,640	1,877
Empennage Group	472	590	444	623
Fuselage Group	3,267	2,149	2,850	1,084
Engine Section	29	0	44	157
Land. Gear Group	986	764	1,382	466
Nose Gear	193	155	254	gears on
Main Gear	793	609	1,128	four fins
Structure Total	6,802	5,197	7,360	4,207
Engine(s)	2,942*	2,254	2,787	2,935
Fuel System	920	362	1,520	185
Propeller Inst.	0	0		1,937
Propulsion System	261	198	396	470
Power Plant Total	4,123	2,814	4,703	5,527
Avionics + Instrum.	54	29	155	64
Surface Controls	509	672	686	364
Hydraulic System	307	406	210	214
Electrical System	662	408	488	377
Electronics	218	75	286	120
Test instrumentation				428
Ballast	337			230
Armament	479		1,008	
Guns or cannons	750		639	97
Air Cond. System	87	65	170	76
Furnishings	184	108	207	196
Miscellanous		30(paint)		121
Fixed Equipm't Total	3,587	1,793	3,849	2,287
$W_{oil} + W_{tfo}$	44	N.A.	50	102
Engine oil	75	23	18	94
Max. Fuel Capacity	4,404	4,440	10,593	1,839
Payload	829(ammo)	N.A.	863(ammo)	0
Crew	230	230	230	200

*Includes afterburners **Delta wing configuration
***F-86 modified with NACA flush side inlets
****Counter-rotating propeller driven tailsitter (VTOL)

Table A13.3b Group Weight Data for NASA X Airplanes

Type	Mc Donnell XF-88A	Convair XF-92A	NAA YF-93A	Convair XFY-1
Flight Design Gross Weight, GW, lbs	20,098	11,600	21,846	14,250
Structure/GW	0.338	0.448	0.337	0.295
Power Plant/GW	0.205	0.243	0.215	0.388
Fixed Equipm't/GW	0.178	0.155	0.176	0.160
Empty Weight/GW	0.722	0.845	0.728	0.844
Wing Group/GW	0.102	0.146	0.121	0.132
Empenn. Group/GW	0.023	0.051	0.020	0.044
Fuselage Group/GW	0.163	0.185	0.130	0.076
Engine Section/GW	0.001	0.000	0.002	0.011
Land. Gear Group/GW	0.049	0.066	0.064	0.033
Take-off Gross Wht, W_{TO}, lbs	20,098	11,600	27,788	15,185
Empty Weight, W_E, lbs	14,512	9,804	15,912	12,021
Wing Group/S, psf	5.9	4.0	8.6	5.3
Emp. Grp/S_{emp}, psf	4.3	7.8	5.6	3.5
Ultimate Load Factor, g's	11.0	11.0	11.0	11.3
Surface Areas, ft^2				
Wing, S	350	425	306	355
Horiz. Tail, S_h	N.A.	0	N.A.	N.A.
Vert. Tail, S_v	N.A.	76.0	N.A.	N.A.
Empenn. Area, S_{emp}	109	76.0	79.6	176

Table A13.4a Group Weight Data for NASA X Airplanes

Type	Lockheed XV-4A*	Lockheed XV-4B**	Ryan XV-5A***	**** Bell X-22A
Number of engines:	2	6	4	4
Weight Item, lbs	Jet	Jet	Jet + liftfan	Turbo-shaft
Wing Group	350	395	1,059	789
Empennage Group	170	167	267	131
Fuselage Group	1,207	1,274	1,341	1,324
Engine Section	245	333	45	610
Land. Gear Group	291	389	482	432
Nose Gear	57	77	82	94
Main Gear	234	312	400	338
Structure Total	2,263	2,558	3,194	3,286
Engine(s) (main)	872	758	913	1,191
Engine(s) (lift)	0	1,500	1,855	
Exhaust system	520	608	304	8
Fuel System	155	142	124	175
Propeller Inst. including drives:				2,147
Ducts and supports: fwd 695, aft 686, total:				1,381
Propulsion System	101	88	80	246
Power Plant Total	1,648	3,096	3,276	5,148
Avionics + Instrum.	73	133	73	121
Surface Controls	486	655	440	1,256
Hydraulic System	62	116	115	162
Electrical System	376	394	196	376
Electronics	29	35	40	237
Test instrumentation	583	200	515	1,520
Auxiliary gear	52	27	158	10
Air Cond. System	32	58	34	45
Furnishings	209	391	235	376
Fixed Equipm't Total	1,902	2,009	1,806	4,103
$W_{oil} + W_{tfo}$	40	30	29	72
Engine oil	20	62	12	22
Max. Fuel Capacity	1,147	3,815	2,430	2,031
Crew	180	430	180	360
Payload				1,200

*Ejector type VTOL **Lift engine type VTOL Liftfan research airplane ****Tiltrotor research airplane

Table A13.4b Group Weight Data for NASA X Airplanes

Type	Lockheed XV-4A	Lockheed XV-4B	Ryan XV-5A	Bell X-22A
Flight Design Gross Weight, GW, lbs	7,200	12,000	9,200	14,700
Structure/GW	0.314	0.213	0.347	0.224
Power Plant/GW	0.229	0.258	0.356	0.350
Fixed Equipm't/GW	0.264	0.167	0.196	0.279
Empty Weight/GW	0.807	0.639	0.900	0.853
Wing Group/GW	0.049	0.033	0.115	0.054
Empenn. Group/GW	0.024	0.014	0.029	0.009
Fuselage Group/GW	0.168	0.106	0.146	0.090
Engine Section/GW	0.034	0.028	0.005	0.041
Land. Gear Group/GW	0.040	0.032	0.052	0.029
Take-off Gross Wht, W_{TO}, lbs	7,200	12,000	9,972	14,700
Empty Weight, W_E, lbs	5,813	7,663	8,276	12,537
Wing Group/S, psf	3.4	3.8	4.1	4.9
Emp. Grp/S_{emp}, psf	3.2	3.1	2.6	1.5
Ultimate Load Factor, g's	7.5	4.5	6.0	4.5
Surface Areas, ft^2				
Wing, S	104	104	260	160
Horiz. Tail, S_h	26.4	26.4	52.9	20
Vert. Tail, S_v	27.5	27.5	51.0	68.5
Empenn. Area, S_{emp}	53.9	53.9	104	88.5

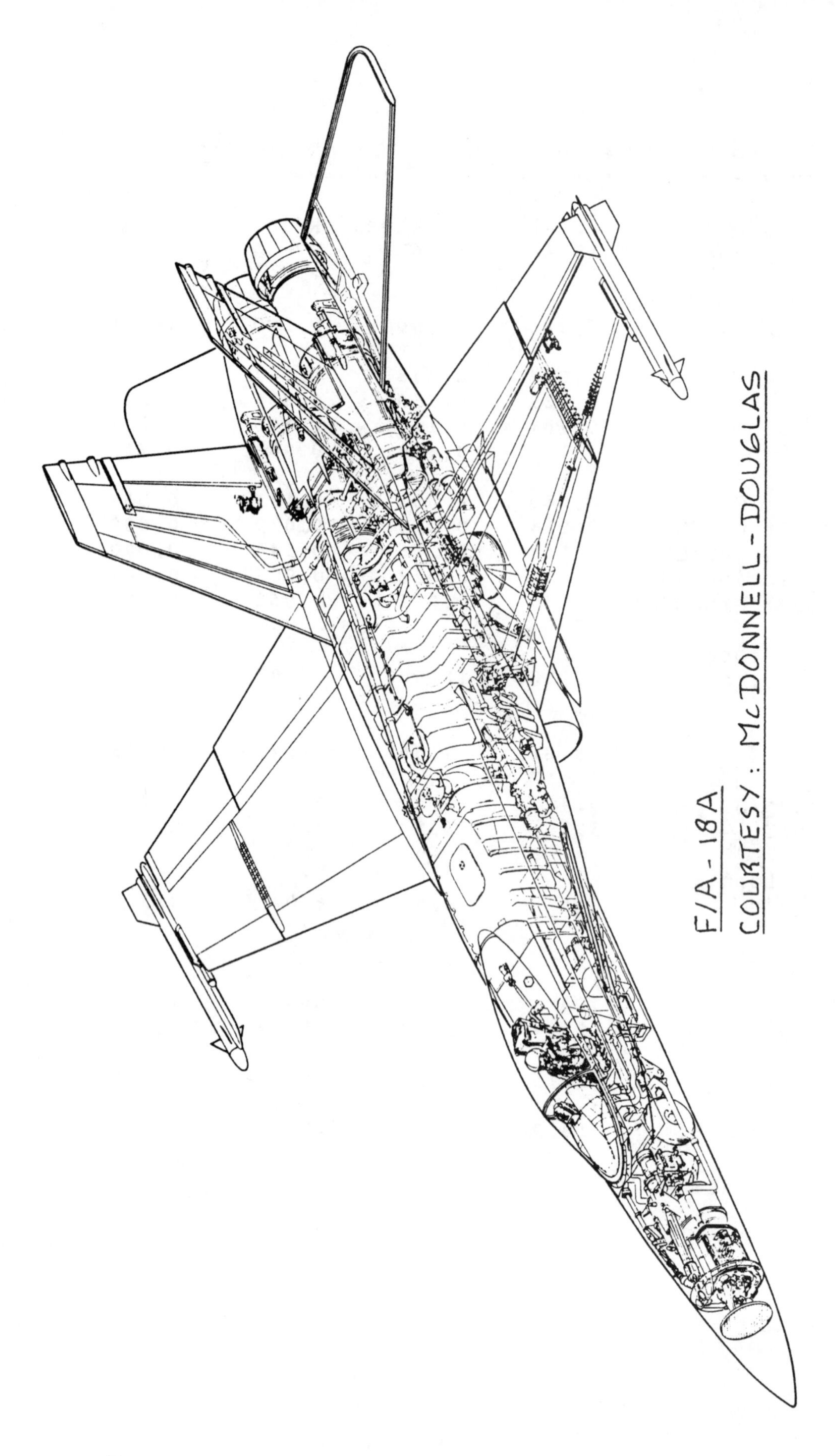
F/A-18A
COURTESY: McDONNELL-DOUGLAS

APPENDIX B: DATA SOURCE FOR NON-DIMENSIONAL RADII OF GYRATION FOR AIRPLANES

The purpose of this appendix is to present tabulated data for non-dimensional radii of gyration of airplanes. Actual moments of inertia can be estimated from these non-dimensional radii of gyration with the help of Equations 3.7 through 3.8.

The tables are organized as follows:

Table B1: Homebuilt propeller driven airplanes
Table B2: Single engine propeller driven airplanes
Table B3: Twin engine propeller driven airplanes
Table B4: Agricultural airplanes
Table B5: Business jets
Table B6: Regional turbopropeller driven airplanes
Table B7a: Jet transports
Table B7b: Piston-propeller driven transports
Table B7c: Turbopropeller driven transports
Table B8: Military trainers
Table B9a: Fighters (Jet)
Table B9b: Fighters (Propeller)
Table B10a: Bombers (Piston-Propeller)
Table B10b: Bombers (Jet)
Table B10c: Military patrol airplanes (Propeller)
Table B10d: Military transports (Propeller)
Table B11: Flying boats
Table B12: Supersonic cruise airplanes

The data in all these table were derived from manufacturers data and/or from Ref.11.

Table B1 Non-dimensional Radii of Gyration for Homebuilt Propeller Driven Airplanes

Airplane Type	GW lbs	Wing Span, b, ft	Total Length, L, ft	e = (b+L)/2, ft	$\bar{R}_x$	$\bar{R}_y$	$\bar{R}_z$	Number of engines and disposition

At the time of printing, no data were available for this type airplane

Table B2 Non-dimensional Radii of Gyration for Single Engine Propeller Driven Airplanes

Airplane Type	GW lbs	Wing Span, b, ft	Total Length, L, ft	e = (b+L)/2, ft	$\bar{R}_x$	$\bar{R}_y$	$\bar{R}_z$	Number of engines and disposition
Beech N-35*	3,125	32.8	25.1	29.0	0.248	0.338	0.393	1 in fusel.
Cessna 150M**	1,127	33.5	21.5	27.5	0.254	0.405	0.418	1 in fusel.
Cessna 172M**	1,477	36.2	26.5	31.4	0.242	0.386	0.403	1 in fusel.
Cessna 177A**	1,761	35.6	27.0	31.3	0.212	0.362	0.394	1 in fusel.
Cessna R182**	1,885	36.2	28.0	32.1	0.342	0.397	0.393	1 in fusel.
Cessna 210K***	2,700	36.8	28.3	32.6	0.222	0.356	0.379	1 in fusel.

*at W_{TO} **at W_{OE} ***atW_{OE} plus 25 percent fuel

Note: one pilot included in all data

Table B3 Non-dimensional Radii of Gyration for Twin Engine

Propeller Driven Airplanes

Airplane Type	GW lbs	Wing Span, b, ft	Total Length, L, ft	e = (b+L)/2, ft	$\bar{R}_x$	$\bar{R}_y$	$\bar{R}_z$	Number of engines and disposition
Beech 55	4,880	37.8	25.7	31.8	0.260	0.329	0.399	2 on wing
Beech 95	4,000	37.8	25.3	31.6	0.251	0.327	0.391	2 on wing
Beech D-50	6,500	45.9	31.5	38.7	0.240	0.313	0.384	2 on wing
Beech D18S	9,000	47.7	34.2	41.0	0.232	0.360	0.396	2 on wing
Cessna 402*	5,000	39.9	36.3	38.1	0.414	0.278	0.502	2 on wing
Cessna 402	6,200	39.9	36.3	38.1	0.373	0.269	0.461	2 on wing
Cessna 404*	4,851	46.7	39.5	43.1	0.324	0.318	0.446	2 on wing
Cessna 404	8,400	46.7	39.5	43.1	0.340	0.284	0.445	2 on wing
Cessna 441*	5,642	49.3	39.0	44.2	0.285	0.345	0.429	2 on wing
Cessna 441	9,925	49.3	39.0	44.2	0.256	0.212	0.336	2 on wing

*at W_E

Table B4 Non-dimensional Radii of Gyration for Agricultural Airplanes

Airplane Type	GW lbs	Wing Span, b, ft	Total Length, L, ft	e = (b+L)/2, ft	$\bar{R}_x$	$\bar{R}_y$	$\bar{R}_z$	Number of engines and disposition

At the time of printing, no data were available for this type of airplane

Table B5 Non-dimensional Radii of Gyration for Business Jets

Airplane Type	GW lbs	Wing Span, b, ft	Total Length, L, ft	e = (b+L)/2, ft	$\bar{R}_x$	$\bar{R}_y$	$\bar{R}_z$	Number of engines and disposition
Morane/S 760	7,066	33.3	32.9	33.1	0.374	0.328	0.486	2 in W/F
Lockh. Jetstar	39,288	53.7	58.8	56.3	0.370	0.356	0.503	4 on fusel.
Cessna 500*	6,505	47.1	43.5	45.3	0.236	0.384	0.430	2 on fusel.
Cessna 500**	12,000	47.1	43.5	45.3	0.306	0.303	0.423	2 on fusel.
Cessna 550*	7,036	51.7	47.2	49.5	0.243	0.400	0.447	2 on fusel.
Cessna 550**	13,500	51.7	47.2	49.5	0.293	0.312	0.420	2 on fusel.

*at W_e **at W_{TO}

Table B6 Non-dimensional Radii of Gyration for Regional Turbopropeller Driven Airplanes

Airplane Type	GW lbs	Wing Span, b, ft	Total Length, L, ft	e = (b+L)/2, ft	$\bar{R}_x$	$\bar{R}_y$	$\bar{R}_z$	Number of engines and disposition
Fokker F-27A	38,500	95.2	77.2	86.2	0.235	0.363	0.416	2 on wing
DHC6 Twin Otter	12,500	65.0	51.8	58.4	0.203	0.326	0.350	2 on wing

Table B7a Non-dimensional Radii of Gyration for Jet Transports

Airplane Type	GW lbs	Wing Span, b, ft	Total Length, L, ft	e = (b+L)/2, ft	$\bar{R}_x$	$\bar{R}_y$	$\bar{R}_z$	Number of engines and disposition
Convair 880	185,000	120.0	124.2	122.1	0.320	0.342	0.465	4 on wing
Convair 880	191,500	120.0	124.2	122.1	0.322	0.339	0.464	4 on wing
Convair 990	240,000	120.0	134.8	127.4	0.335	0.338	0.473	4 on wing
Convair 990	245,000	120.0	134.8	127.4	0.305	0.334	0.472	4 on wing
Boeing 727-100	165,000	108.0	133.2	120.6	0.249	0.375	0.452	3 on fusel.
Boeing 727-100*	89,000	108.0	133.2	120.6	0.247	0.442	0.518	3 on fusel.
Boeing 727-200	180,000	108.0	153.2	130.6	0.248	0.394	0.502	3 on fusel.
Boeing 727-200*	100,000	108.0	153.2	130.6	0.240	0.451	0.550	3 on fusel.
Boeing 737-200	113,000	93.0	100.0	96.5	0.246	0.382	0.456	2 on wing
Boeing 737-200*	62,000	93.0	100.0	96.5	0.264	0.456	0.517	2 on wing
Boeing 747-100B	800,000	195.7	231.3	213.5	0.290	0.329	0.445	4 on wing
Boeing 747-100B*	350,000	195.7	231.3	213.5	0.332	0.380	0.508	4 on wing
McDD DC9-10	74,000	89.4	104.3	96.9	0.242	0.360	0.435	2 on fusel.
McDD DC8	210,000	142.4	150.5	146.5	0.301	0.349	0.434	4 on wing

*at W_{OE}

Table B7b Non-dimensional Radii of Gyration for Piston-Propeller Driven Transports

Airplane Type	GW lbs	Wing Span, b, ft	Total Length, L, ft	e = (b+L)/2, ft	$\bar{R}_x$	$\bar{R}_y$	$\bar{R}_z$	Number of engines and disposition
Lockheed L-749A	107,000	123.0	95.2	109.1	0.300	0.298	0.426	4 on wing
Lockheed L-1049	120,000	123.0	113.6	118.3	0.316	0.336	0.448	4 on wing
Lockheed L-1649	146,500	150.0	116.2	133.1	0.371	0.278	0.473	4 on wing
Douglas DC-4	60,360	138.3	97.6	118.0	0.250	0.320	0.388	4 on wing
Douglas DC-6	97,200	117.5	100.5	109.0	0.322	0.324	0.456	4 on wing
Airspeed Ambass.	49,500	115.0	80.4	97.7	0.278	0.314	0.400	2 on wing
Martin 404	45,000	93.3	74.6	84.2	0.272	0.378	0.444	2 on wing
Convair T-240	41,800	91.7	74.7	83.2	0.286	0.351	0.443	2 on wing
Convair T-340	44,500	105.7	79.2	92.4	0.308	0.345	0.457	2 on wing
Beech Twin Quad	20,000	70.0	52.7	61.4	0.225	0.303	0.346	4 in wing

Table B7c Non-dimensional Radii of Gyration for Turbo-Propeller Driven Transports

Airplane Type	GW lbs	Wing Span, b, ft	Total Length, L, ft	e = (b+L)/2, ft	$\bar{R}_x$	$\bar{R}_y$	$\bar{R}_z$	Number of engines and disposition
Bristol 175(k)*	103,000	130.0	110.0	120.0	0.317	0.356	0.455	4 on wing
Bristol 167(1)**	187,000	230.0	177.0	203.5	0.330	0.356	0.478	4 on wing
Lockh. Electra	116,000	99.0	104.7	101.9	0.394	0.341	0.497	4 on wing

*Britannia **Brabazon

Table B8 Non-dimensional Radii of Gyration for Military Trainers

Airplane Type	GW lbs	Wing Span, b, ft	Total Length, L, ft	e = (b+L)/2 ft	$\bar{R}_x$	$\bar{R}_y$	$\bar{R}_z$	Number of engines and disposition
Cessna T-37A	6,300	38.4	30.0	34.2	0.220	0.142	0.245	2 in fusel.

Table B9a Non-dimensional Radii of Gyration for Fighters (Jet)

Airplane Type	GW lbs	Wing Span, b, ft	Total Length, L, ft	e = (b+L)/2, ft	$\bar{R}_x$	$\bar{R}_y$	$\bar{R}_z$	Number of engines and disposition
McD F2H-1	14,413	41.6	40.2	40.9	0.230	0.359	0.465	2 in W/F
McD F3H-2N	26,878	35.3	58.8	47.1	0.252	0.107	0.449	1 in fusel.
McD F-101A	36,969	39.7	67.4	53.6	0.209	0.329	0.428	2 in fusel.
VS Attacker	10,450	36.9	37.3	37.1	0.244	0.328	0.400	1 in fusel.
DH Vampire 20	10,891	40.0	30.1	35.1	0.286	0.318	0.409	1 in fusel.
Gl. Meteor II	11,100	43.0	41.4	42.2	0.286	0.330	0.404	2 in wing
Lockheed F-80A	11,940	38.9	34.3	36.6	0.286	0.356	0.444	1 in fusel.
Lockheed F-94B	13,650	37.5	40.1	38.8	0.284	0.396	0.488	1 in fusel.
Lockheed F-104G	20,900	21.9	54.8	38.4	0.224	0.392	0.563	1 in fusel.
NAA F-86A	13,900	37.1	37.5	37.3	0.266	0.346	0.400	1 in fusel.
NAA FJ-3	16,883	37.0	37.6	37.3	0.281	0.352	0.438	1 in fusel.
NAA F-100D	29,800	38.0	47.0	42.5	0.252	0.376	0.462	1 in fusel.
Vought XF8U-1	21,300	35.7	54.4	45.1	0.225	0.404	0.507	1 in fusel.
Vought F8U-3	30,600	40.0	58.9	49.5	0.225	0.375	0.467	1 in fusel.
GD XF-91	18,600	31.3	43.3	37.3	0.323	0.424	0.548	1 in fusel.
GD TF-102A	32,859	38.1	63.2	50.7	0.295	0.386	0.520	1 in fusel.
GD F-106B	36,834	38.3	70.7	54.5	0.247	0.379	0.516	1 in fusel.
Northrop F-89D	38,000	58.0	54.0	56.0	0.440	0.304	0.532	2 in fusel.
Republic RF-84F	19,000	33.6	47.5	40.6	0.310	0.308	0.432	1 in fusel.
Republic F-105D	34,058	35.0	64.4	49.7	0.231	0.425	0.567	1 in fusel.
Grumman F9F-8	16,744	34.5	41.9	38.2	0.248	0.374	0.454	1 in fusel.
Grumman XF10F-1	26,160	36.8	57.8	46.9	0.251	0.323	0.414	1 in fusel.
Grumman F11F-1	16,500	31.6	40.8	36.2	0.221	0.404	0.484	1 in fusel.

Table B9b Non-dimensional Radii of Gyration for Fighters (Propeller)

Airplane Type	GW lbs	Wing Span, b, ft	Total Length, L, ft	e = (b+L)/2, ft	$\bar{R}_x$	$\bar{R}_y$	$\bar{R}_z$	Number of engines and disposition
Brewster Buffalo	5,066	35.0	26.0	30.5	0.208	0.358	0.374	1 in fusel.
Seversky P35	5,788	36.0	26.8	31.4	0.198	0.367	0.360	1 in fusel.
VS Spitfire-I	6,250	36.8	29.9	33.4	0.240	0.334	0.384	1 in fusel.
BP Defiant	6,410	39.4	35.0	37.2	0.234	0.360	0.404	1 in fusel.
Curtiss P36	6,825	37.3	31.7	34.5	0.172	0.356	0.370	1 in fusel.
Bell P39	7,533	34.0	30.0	32.0	0.276	0.340	0.425	1 in fusel.
Grumman F6F	10,560	42.8	33.5	38.2	0.242	0.346	0.404	1 in fusel.
Hawker Typhoon	11,017	41.5	31.7	36.6	0.277	0.300	0.394	1 in fusel.
Republic P47	12,500	40.7	36.0	38.4	0.296	0.322	0.428	1 in fusel.
Vought F4U	12,850	41.0	34.5	37.8	0.268	0.360	0.420	1 in fusel.
Bl.Firebr'd-III	13,660	49.8	38.2	44.0	0.250	0.300	0.397	1 in fusel.
Westland Welkin	18,340	70.0	42.0	56.0	0.270	0.304	0.408	2 in wing
Bristol Beauftr	22,635	57.8	42.5	50.2	0.330	0.299	0.447	2 in wing
Bristol Brigand	39,000	71.7	46.4	59.1	0.299	0.338	0.438	2 in wing

Table B10a Non-dimensional Radii of Gyration for Bombers (Piston-Propeller)

Airplane Type	GW lbs	Wing Span, b, ft	Total Length, L, ft	e = (b+L)/2, ft	$\bar{R}_x$	$\bar{R}_y$	$\bar{R}_z$	Number of engines and disposition
Martin B-26	26,600	65.0	57.6	61.3	0.270	0.320	0.410	2 on wing
HP Halifax	55,000	99.0	71.6	85.3	0.346	0.306	0.395	4 on wing
Shorts Stirling	64,000	99.0	87.3	93.2	0.360	0.330	0.488	4 on wing
Boeing B-29	105,000	141.0	99.0	120.0	0.316	0.320	0.376	4 on wing
Boeing B-50	120,000	141.0	99.0	120.0	0.304	0.332	0.450	4 on wing
GD B-36	357,500	230.0	162.0	196.0	0.316	0.262	0.428	6 in wing

Table B10b Non-dimensional Radii of Gyration for Bombers (Jet)

Airplane Type	GW lbs	Wing Span, b, ft	Total Length, L, ft	e = (b+L)/2, ft	$\bar{R}_x$	$\bar{R}_y$	$\bar{R}_z$	Number of engines and disposition
Martin XB-51	53,785	53.0	81.0	67.0	0.194	0.404	0.498	3 in/on fus.
Martin B57A	48,554	64.0	66.0	65.0	0.312	0.278	0.412	2 in wing
Boeing XB-47	125,000	116.0	107.0	111.5	0.346	0.320	0.474	4 in wing
Boeing B-52A	390,000	185.0	156.5	170.8	0.346	0.306	0.466	8 on wing
Northrop RB-49A	213,500	172.0	53.0*	112.5	0.316	0.316	0.510	6 in wing
NAA B-45A	82,600	89.0	75.3	82.2	0.325	0.290	0.438	4 on wing
NAA B-45C	82,600	89.0	75.3	82.2	0.340	0.299	0.456	4 on wing

*Flying wing

Table B10c Non-dimensional Radii of Gyration for Military Patrol Airplanes

(Propeller Driven)

Airplane Type	GW lbs	Wing Span, b, ft	Total Length, L, ft	e = (b+L)/2, ft	$\bar{R}_x$	$\bar{R}_y$	$\bar{R}_z$	Number of engines and disposition
Piston-Propeller Driven								
Lockheed P2V-4	67,500	100.0	81.6	90.8	0.368	0.300	0.484	2 on wing
Lockheed P2V-7*	67,500	100.0	91.7	95.9	0.372	0.266	0.462	4 on wing
Grumman S2F-3	26,147	69.7	43.5	56.6	0.240	0.347	0.387	2 on wing
Grumman W2F-1	41,549	80.6	56.3	68.4	0.235	0.366	0.387	2 on wing
Turbo-propeller Driven								
Lockheed P3V-1	127,200	99.7	116.8	108.3	0.357	0.249	0.421	4 on wing

Table B10d Non-dimensional Radii of Gyration for Military Transports

(Propeller Driven)

Airplane Type	GW lbs	Wing Span, b, ft	Total Length, L, ft	e = (b+L)/2, ft	$\bar{R}_x$	$\bar{R}_y$	$\bar{R}_z$	Number of engines and disposition
Piston-Propeller Driven								
Douglas C-54	61,840	117.5	94.0	105.8	0.286	0.294	0.406	4 on wing
Fairchild C-119B in wing	64,000	109.3	88.5	98.9	0.287	0.282	0.390	2 0n wing
Boeing C-97	128,340	141.2	110.3	125.8	0.276	0.325	0.424	4 on wing
GD XC-99	265,000	230.0	182.5	206.3	0.276	0.346	0.432	6 on wing
Turbo-propeller Driven								
Lockheed C-130B	135,000	132.6	97.8	115.2	0.363	0.319	0.489	4 on wing
Lockheed C-130E	155,000	132.6	97.8	115.2	0.375	o.290	0.486	4 on wing

*Has two jet engines outboard of the piston engines

Table B11 Non-dimensional Radii of Gyration for Flying Boats

Airplane Type	GW lbs	Wing Span, b, ft	Total Length, L, ft	e = (b+L)/2, ft	$\bar{R}_x$	$\bar{R}_y$	$\bar{R}_z$	Number of engines and disposition
XBTM-1	20,009	50.0	41.2	45.6	0.230	0.340	0.380	1 in fusel.
XP4M-1	80,000	114.0	84.0	99.0	0.248	0.320	0.414	2 on wing
VS Seagull	14,230	52.5	44.0	48.3	0.297	0.364	0.402	1 in WF

Table B12 Non-dimensional Radii of Gyration for Supersonic Cruise Airplanes

Airplane Type	GW lbs	Wing Span, b, ft	Total Length, L, ft	e = (b+L)/2, ft	R_x	R_y	R_z	Number of engines and disposition
NAA A3J-1	44,305	53.0	72.5	62.8	0.240	0.372	0.472	2 in fusel.

12. INDEX

Notes

Notes

Notes

Notes

Notes